**Chandrashekar Natarajan,
Lee Hales**

Simplified Systematic Network Planning

Chandrashekar Natarajan,
Lee Hales

Simplified Systematic Network Planning

Six Steps to Effective Network Planning

VDM Verlag Dr. Müller

Imprint

Bibliographic information by the German National Library: The German National Library lists this publication at the German National Bibliography; detailed bibliographic information is available on the Internet at
http://dnb.d-nb.de.

Any brand names and product names mentioned in this book are subject to trademark, brand or patent protection and are trademarks or registered trademarks of their respective holders. The use of brand names, product names, common names, trade names, product descriptions etc. even without a particular marking in this works is in no way to be construed to mean that such names may be regarded as unrestricted in respect of trademark and brand protection legislation and could thus be used by anyone.

Cover image: www.purestockx.com

Published 2008 Saarbrücken

Publisher:
VDM Verlag Dr. Müller Aktiengesellschaft & Co. KG , Dudweiler Landstr. 125 a,
66123 Saarbrücken, Germany,
Phone +49 681 9100-698, Fax +49 681 9100-988,
Email: info@vdm-verlag.de

Produced in Germany by:
Reha GmbH, Dudweilerstrasse 72, D-66111 Saarbrücken
Schaltungsdienst Lange o.H.G., Zehrensdorfer Str. 11, 12277 Berlin, Germany
Books on Demand GmbH, Gutenbergring 53, 22848 Norderstedt, Germany

Impressum

Bibliografische Information der Deutschen Nationalbibliothek: Die Deutsche Nationalbibliothek verzeichnet diese Publikation in der Deutschen Nationalbibliografie; detaillierte bibliografische Daten sind im Internet über http://dnb.d-nb.de abrufbar.

Alle in diesem Buch genannten Marken und Produktnamen unterliegen warenzeichen-, marken- oder patentrechtlichem Schutz bzw. sind Warenzeichen oder eingetragene Warenzeichen der jeweiligen Inhaber. Die Wiedergabe von Marken, Produktnamen, Gebrauchsnamen, Handelsnamen, Warenbezeichnungen u.s.w. in diesem Werk berechtigt auch ohne besondere Kennzeichnung nicht zu der Annahme, dass solche Namen im Sinne der Warenzeichen- und Markenschutzgesetzgebung als frei zu betrachten wären und daher von jedermann benutzt werden dürften.

Coverbild: www.purestockx.com

Erscheinungsjahr: 2008
Erscheinungsort: Saarbrücken

Verlag: VDM Verlag Dr. Müller Aktiengesellschaft & Co. KG , Dudweiler Landstr. 125 a,
D- 66123 Saarbrücken,
Telefon +49 681 9100-698, Telefax +49 681 9100-988,
Email: info@vdm-verlag.de

Herstellung in Deutschland:
Schaltungsdienst Lange o.H.G., Zehrensdorfer Str. 11, D-12277 Berlin
Books on Demand GmbH, Gutenbergring 53, D-22848 Norderstedt
Reha GmbH, Dudweilerstrasse 72, D-66111 Saarbrücken

ISBN: 978-3-639-02546-0

SIMPLIFIED SYSTEMATIC NETWORK PLANNING

Developed and Presented by

Chandra S. Natarajan, PMP, CPIM, CSCP, CLP, CSSBB, CPM

Chandra Natarajan is a Sr. Supply Chain Manager with The Pepsi Bottling Group. He has network modeling experience with simple tools such as Excel Solver, AMPL, and LINDO, and advanced tools from JDA, SAP, Manhattan, and i2. Mr. Natarajan has been a speaker on network modeling for the JDA Client Steering Committee and the Georgia Tech Supply Chain Logistics Institute. He holds a Bachelor's in Mechanical Engineering from Jawaharlal Nehru Technological University, India and an MS in Industrial & Systems Engineering from Georgia Tech.

H. Lee Hales

Lee Hales is President of Richard Muther & Associates industrial management consultants. He is a member of CSCMP and has worked in distribution planning and site location for more than 30 years. Mr. Hales is also a Senior Lecturer for the Georgia Tech Supply Chain Logistics Institute and a past director of the Institute of Industrial Engineers' Facilities Planning & Design Division. He holds BA and MA degrees from the University of Kansas and an MS from the Sloan School, Massachusetts Institute of Technology.

First edition - March 2008

ISBN 0-933684-21-5
ISBN-13: 978-0-933684-21-8

Copyright © 2008 by Chandra S. Natarajan and Richard Muther & Associates. Printed in the United States of America. All rights reserved. This book, or parts thereof, may not be reproduced in any form without expressed permission in writing of the holder of the copyright. The green forms in the back are excepted; readers are encouraged to reproduce and use them for their personal or in-company need, provided the original source is not deleted.

Published by: Management and Industrial Research Publications
4129 River Cliff Chase, Marietta, GA 30067
Tel: (770) 798-7792; Fax: (770) 859-0166
www.MIRPBooks.com

Other booklets in this series:

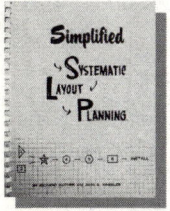

Simplified Systematic Layout Planning

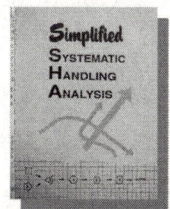

Simplified Systematic Handling Analysis

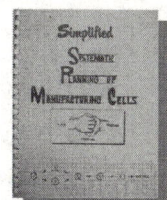

Simplified Systematic Planning of Manufacturing Cells

Dick Ward
Material Handling Industry of America (ret.)

"As an engineering consultant, educator and industry practitioner I have promoted and personally used the series of systematic planning books and short courses authored and presented by the colleagues of Richard Muther Associates for more than 40 years. When it comes to being practical and the principles easily applied, there are non better! And now we have the latest in the series, Simplified Systematic Network Planning, which brings the planning process of what are really complex issues down to a level where even the most novice among us can benefit. Authors Chandra Natarajan and Lee Hales bring industrial experience, a presentation format and a planning pattern that is second to none. If you want to hit the ground running on your next (or your first) network planning assignment, start by reading Simplified Systematic Network Planning by Natarajan and Hales."

John M. Hill
TranSystems | ESYNC

"Excellent introduction to optimization with rich illustrations and examples. Practical discussion of the fundamentals of network analysis throughout the book, coupled with a step-by-step measured approach to execution."

David B. Scott
TranSystems | ESYNC

"The document preface describes the first-time network planner as the primary beneficiary of the comprehensive network planning methodology defined in The Simplified Systematic Network Planning book. As an experienced practitioner of network planning, I found the rigorous methodology and documentation templates in the book are invaluable in bringing order to a complex process".

TABLE OF CONTENTS

Preface

About Network Planning & Modeling

Network Planning Terminology

Introduction to Simplified SNP

- 1 — ORIENT the project

- 2 — DEFINE the variables

- 3 — ANALYZE the sensitivities

- 4 — CREATE scenarios

- 5 — EVALUATE the alternatives

- 6 — DETAIL and DO

Case Examples

Working forms

Synopsis of full SNP

PREFACE

This booklet is written chiefly for three groups:

- The planner of logistical networks
- Team members involved in a network planning project
- The team leader, teacher, or facilitator for those doing the planning

This booklet presents an organized, proven method – always helpful. We have found this process to be highly effective for first-time network planners, for those invited to be involved in the planning but don't know how, and for those who may be skilled at applying network analysis tools but need a clear guide on how to successfully complete a network planning project.

This booklet includes six specific steps. It tells how to conduct a successful network planning project. It provides a carry-through case example. It gives you working forms to use. And it gives numerous other explanations and specific cases.

Most network planning involves modeling. You cannot model logistical networks without software – often a specialized program using some form optimization software. We assume that you have such software and know how to use it. This booklet and method address the questions of how to organize, plan and conduct network planning projects in which modeling software will be used.

Simplified Systematic Network Planning (Simplified SNP) is especially helpful when:

- Network planning is new for you or your company and you need a standard process.
- Your project is small or a sub-project of a larger network plan.
- You want faster project start-up with less confusion on purpose, scope and approach.
- You want to finish on time or early without delays, excessive iterations and rework.
- You want visualization of the network being modeled, with clear and explicit diagrams.
- You want accurate modeling that predicts savings close to actual results.
- You want effective involvement of operating personnel and acceptance of recommendations.
- You want effective communication and documentation of the network model and outputs.
- You want to present several viable alternatives to your management.
- You want to choose your network plan based on all factors, economic and intangible.

Systematic Network Planning (SNP) is based upon the High Performance Planning model developed by Richard Muther. As such, it represents an extension of his work and an application of his latest general purpose planning method: Planning by Design (PxD). Dr. Richard Ward, a leading educator in the fields of material handling and logistics, reviewed our method during its development and made helpful suggestions for our method and this book. We are grateful to both.

We sincerely hope this booklet will be of direct help to you and your associates.

                                                      Chandra S. Natarajan
                                                      PMP, CPIM, CSCP, CLP, CSSBB

                                                      H. Lee Hales
                                                      President
                                                      Richard Muther & Associates

ABOUT NETWORK PLANNING & MODELING

Network Planning

Supply chain and logistical networks consist of locations – suppliers, plants, warehouses, and customers – and the transportation routes or lanes between them. Planning such networks typically requires the following decisions:

- How many plants and warehouses should we have?
- Which customers will be served and from which locations?
- Which products will be made internally and which sourced from outside?
- Which suppliers will be used and for what items?
- Which products will be made or distributed at which locations, and in what quantities?
- How much capacity will be provided at each producing or distributing location?
- How much inventory will be held at which locations?
- What will be the hours and days of operation?
- What modes of transportation will be used between locations?

Network planning seeks to maximize the company's profits or minimize costs subject to relevant constraints and important intangible considerations. Planning also validates future strategies and prepares for various contingencies.

The cost of poor network decisions can be high and generally increases as companies grow, more products are added, and their networks become more complex.

Network Modeling

Even in small networks with a few locations and routes, sound decisions require planners to gather large amounts of data, investigate numerous variables and evaluate multiple cost trade-offs between warehousing, production, and transportation.

The field of network modeling has evolved to help planners make sound and timely decisions. Using mathematical optimization, planners seek to minimize total costs or maximize profits for relevant lanes and locations while observing appropriate constraints.

Modeling software enables planners to rapidly analyze large amounts of data and to explore many alternatives. Great strides have been made in the development of standardized modeling tools. These are now accessible to many companies and are becoming a part of their day to day decision making process.

But having good software is only the first step to effective network planning. Modeling results and recommendations are only as good as the planner's problem formulation, assumptions, and data. And often, acceptance of the planner's recommendations requires that decision-makers see and understand what was done in the modeling exercise.

Systematic Planning

For best results, network planners should follow a standardized, step-by-step approach. Their planning should be explicit, well-documented and visual. Otherwise, projects will be exposed to re-work and unnecessary iterations, delays, and even rejection of results and recommendations. Given the big savings typically at stake, every week or month of delayed decision and implementation represents costly waste.

"Systematic planning" refers to the way in which planning is structured and conducted, step-by-step. The method presented in this booklet applies the High Performance Planning structure developed by Richard Muther and used in the well-known Muther methods of systematic facilities planning. The procedures described can be mastered in less than one day without formal training in network optimization. However, the planner will still need training in the optimization software necessary to perform network modeling.

NETWORK PLANNING TERMINOLOGY

As you read this booklet you will find specialized terms used by network planners and modelers. If these terms are unfamiliar, the brief definitions below will help you to better understand them. While complete with respect to this booklet, the glossary below is only a partial list of terms used in the field.

Locations	Locations are nodes in a network model. Examples are: suppliers, manufacturing plants, and branch warehouses or distribution centers.
Lanes	Lanes refer to possible transportation routes between locations. Lanes are the arcs that connect the nodes in a network model.
Product Hierarchy	Product hierarchy represents the level of product aggregation in a network model. For example, in the beverage industry, a Stock Keeping Unit (SKU) represents a lower level of the product hierarchy than a package family. Planners must choose the appropriate hierarchy and aggregation for each modeling exercise.
Resources	Resources transform, hold or move products at nodes and arcs in the network model. Resources have capabilities, capacities, and speeds, and costs, and are subject to constraints.
Time Buckets	Time bucket refers to the interval that will be used to describe costs, production and demand. Time buckets are typically days, weeks, or months, depending upon the situation being modeled.
Unit of Measure	Unit of measure is the base scale used for measurement of physical volumes in demand, transportation, production and capacity. Typical units of measures in network planning are units, cases, and pallets.
Model Constraints	Model constraints are limiting conditions applied to demand or resources. An example of a demand constraint would be "all customer demand must be satisfied". An example of a resource constraint would be "plants may not operate more than 120 hours per week".
Model Assumptions	Model assumptions are something accepted as true without proof. Assumptions are made to simplify the modeling effort. They should be acceptable to subject matter experts and decision-makers using the model results.
Model Variables	In network modeling, a variable is any element of the model that can be changed or varied. Thus, "variables" include: design characteristics, parameters, constraints, assumptions, formulas, and even data sources for demand, resources, and costs.
Model Design Characteristics	Design characteristics are the features or elements in a model that distinguish it from other models. They include: Locations, lanes, products and product hierarchy, resources, demand data types and durations, time buckets, and units of measure.
Sourcing	Sourcing refers to a network optimization problem in which modeling identifies locations with the lowest total costs of supply for raw materials or products.
Baseline	A baseline models the current network. A validated baseline gives values that are close to the actual values of the current network, typically measured in terms of costs, production, and demand.
Scenarios	Scenarios represent possible future states of a network. Several scenarios are modeled in the typical network planning exercise. Each is generated by making selected changes to the baseline model.
Sensitivities	The degrees to which model results are affected by changes to model variables.
Variance	Variance refers to the deviation of actual results from expected results.
Typical optimization formula: $\text{Min Cost} \sum_i \sum_j c_{ij} x_{ij}$ st $\sum_i x_{ij} - \sum_k x_{jk} = b_j \; \forall j \in N$	Minimize flow (x) cost (c) between all pairs ($\sum\sum$) of network locations (i's and j's), subject to (st) the conservation constraint that all flows into any location (X_{ij}) must be accounted for, either as outflows (X_{jk}) or as quantities remaining (b_j) at each ($\forall$) location, where j is a type of ($\in$) number N, either integer or real – to require only whole locations or to permit a fractional result.

INTRODUCTION TO SIMPLIFIED SNP

Simplified Systematic Network Planning -- typically termed Simplified SNP -- is a set of six procedures for planning logistical networks. It is suited to smaller projects that do not require the full SNP method.

Basically, every network planning project involves three fundamentals:

1. **Variables** to be defined: In logistical network modeling, a variable is any element of the model that can be changed or varied. Thus, "variables" include: network design characteristics, parameters, constraints, assumptions, formulas, and even data sources for demand, resources, and costs.
2. **Sensitivities** to be analyzed: The degrees to which model results are affected by changes to the model variables.
3. **Scenarios** to be created: Scenarios represent possible future states of a network. Several scenarios are modeled in the typical network planning project. Each is generated by making selected changes to a baseline network model.

The six steps of Simplified SNP follow these three fundamentals and the six steps from a pattern. The pattern may be symbolically indicated, as on the cover of this booklet, and is conceptually drawn here. Each of the six steps carries its own ease-of-recall symbol.

1   **Grid** – the scope of the project and its schedule.

2   **Arrow** – define variables that will drive the analysis.

3   **Bell Curve** – analyze the sensitivities of variables.

4   **Stack of sheets** – scenarios for alternative plans.

5   **Pentagon** – consider all sides; evaluate all factors.

6   **Rectangle** – the implementation plan.

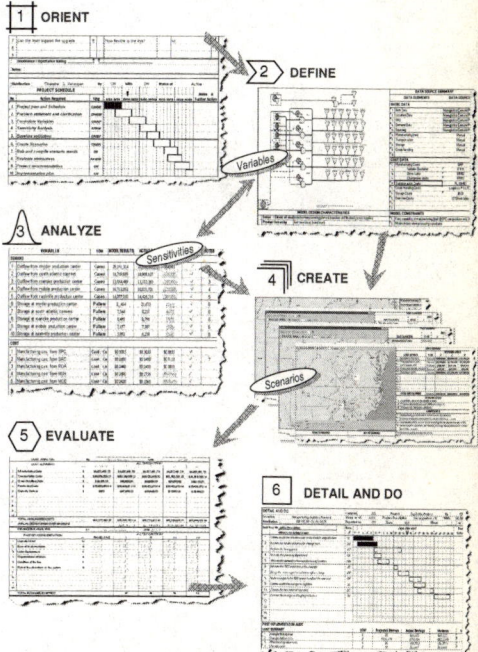

Each of these six steps will be explained in the following chapters. Each chapter follows the same arrangement. A single case example follows through all six steps – on Page 2 of each chapter. Supporting information is shown on subsequent pages of each chapter. At the end of the booklet are three complete six-step *examples*. There also, you will find a blank set of the *working forms*.

The follow-through (or carry-through) example, on Page 2 of each chapter, shows how to perform each step. It involves finding the best existing location at which to upgrade a bottling production line for increased capacity.

Orient t

he Project

STEP 1: ORIENT THE PROJECT

Our first step is to get the project organized.

WHAT YOU DO
- Identify the project, its objective(s) and situation.
- Agree on the scope of the plan and the form of output desired from the planning.
- Document and rate the planning issues.
- Make a plan for modeling and modifying the network.

Key Document = *Orientation & Issues Worksheet*

Output = *The Project Understood and Scheduled*

WHY YOU DO IT

We cannot promise and deliver timely results without a thorough understanding of the situation and a step-by-step plan for our project.

HOW YOU DO IT

1 Complete the heading on the form to identify the worksheet.

2 Determine and fill-in the information regarding the following:
 (a) The objective(s) – or goal(s) — of this project both economic and intangible, and the appropriate objective function for the network model, e.g. minimize relevant costs or maximize profits.
 (b) Implications of not taking action or of making a poor decision.
 (c) Situation & Motivation: choices and factors to consider (costs, productivity, ease of implementation…)
 (d) Scope of the project – How big? How detailed? When needed? Form of planning's output – written report, ROI analysis, plan approved ready for implementation, network re-design.

3 In the Planning Issues section, identify, right up front, the problems that will be faced. Enter each problem, uncertainty, question, matter of dispute or discussion….one line for each.

4 In the two narrow columns, record two things:
 (a) How important the issue is to network planning. Here you record a vowel letter as your order-of magnitude, judgmental rating:
     A – Abnormally important
     E – Especially important
     I – Important issue
     O – Ordinary importance
     U – Unimportant
 (b) If the issue is beyond your control and/or your planning must live with the condition involved, record "X" in the column so you don't waste further time wrestling with that issue.

5 The Action to Resolve the issue and the proposed resolution will be addressed later on in the planning.

6 In the Project Schedule section, list each task, activity, or action required to accomplish the objectives.

7 Shows who is responsible for doing each task.

8 Set a calendar schedule at the top of vertical lines sections, letting each line represent one day. Then schedule each task.

9 There is room at the right for notes or further action regarding each task.

ORIENTATION & ISSUES WORKSHEET ①

Company	MTT
By	CSN
Date	9/23

Project Name	Big Bottle Analysis
With	LH
Sheet	1 of 1

② **Objective(s):** To cost-justify a second, up-graded production line for 32-oz. (big) bottles. To identify the best location for the second line among six existing plants.

Implication: Existing big bottle line at Sommersville is at capacity. Sales will be lost without additional capacity.

Situation & Motivation: Sommersville is the only source of 32 oz. production. Another line is needed but could be added at a different plant location, depending upon costs and other factors. A second location will lower transportation costs.

Scope and Form of Output: Use the current sourcing model built in M2's Strategy software to find the best location. Evaluate intangible considerations as risk and ease of implementation, in addition to optimization of network costs.

③	PLANNING ISSUES	④	⑤ Action to Resolve	Resp	Proposed Resolution	✓ ok
1.	Understanding the new transportation lane costs.	E	Decide how to evaluate costs.	CSN/BF	Meet with Peter	✓
2.	Impact of raw material supplies?	I	Determine if costs change.	CSN/LH	Meet with Debra	✓
3.	What is the cost of conversion at Jonesville?	I	Develop cost per case.	CSN/LH	Meet with Cathy	✓
4.	Are there any backhaul opportunities?	O	Determine potential and impact.	LH	Meet with Peter	✓
5.	Are there any new products to be launched?	O X	Obtain forecast and mktg. plans.	LH	Meet with Michael	✓
6.	How to split the customer demand since 32 oz. only appears on Sommersville PO's?	A	Decide how to allocate demand to plants.	CSN	Meet with Henry	✓
7.	Which bottle lines can be upgraded to 32 oz.?	E	Assess line flexibilities.	CSN	Meet with Don	✓
8.	Bessemer line cannot be upgraded.	A X	Keep it in the model but exclude from alternatives.			
9.						

Dominance / Importance Rating ↑ ↑ Mark "X" if beyond control or scope of project

Notes: 1. - Additional volume available to consume excess big bottle capacity will be analyzed later, after selecting the line location.

Distribution	CSN	By	CN	With	DH	Status of	10/01 Active		
	PROJECT SCHEDULE								
No.	⑥ Action Required	⑦ Who	9/23-9/28	10/01-10/05	10/08-10/12	10/15-10/19	10/2210/26	Notes & Further Schedule	
1.	Project plan and Schedule	CN/DH	■						
2.	Problem statement and clarification	DH/DS			⑧		⑨		
3.	Formulate Variables	CN/KW							
4.	Sensitivity Analysis	KW/BF							
5.	Baseline validation	CN/BF							
6.	Create Scenarios	CN/DS							
7.	Run and compile scenario results	DS							
8.	Evaluate alternatives	KW/BF							
9.	Present recommendation	CN							
10.	Implementation plan	CN							

Notes: 2. - Meeting of subject matter experts to validate the model results against current network performance indicators.

© Copyright 2007. RICHARD MUTHER - 773   May be reproduced for in-company use provided original source is not deleted

... STEP 1 (continued)

There are three sections to the Orientation & Issues Worksheet – our key document in Step 1.

ORIENTATION SECTION

The upper third helps us get on top of the project. It forces us to:
- Describe or define – to clarify what we are doing.
- Record why – the objective(s) we are after.
- Relate our network planning to larger business issues and decisions, helping us "see it whole".
- Identify the situational considerations to be met.
- Question the scope and detail of the network model.
- Clarify what form of output this planning should present.

We are asking ourselves, right at the start of the project, who is going to do the planning, when it is due, the problems expected, and how and when the project will be done. This tends to assure involvement early in the project.

Note: If you find your project is too comprehensive or complex, you may wish to apply full Systematic Network Planning (SNP). Simplified SNP is for small, simple network planning problems. (See the inside of back cover of this booklet for a comparison of Simplified SNP with the full version.)

ISSUES SECTION

An "issue" is a problem or opportunity which affects the planning of the network and/or its subsequent implementation. These must be resolved as part of the planning process. Typical issues include:

> How to allocate costs to locations?
> How to allocate costs to products?
> How to allocate demand to locations, lanes and resources?
> Who will provide the necessary data?
> What are the sources of the data required?
> Understanding the technical constraints on resources being modeled.
> How to model backhaul opportunities?
> How to model new or planned future products lacking historical data?
> Who is available to work on the modeling project?
> Who are the subject matter experts? What are their availabilities?

Getting the right people involved at the outset is the best way to identify issues that are relevant to your project.

PROJECT SCHEDULE

Most projects are more effective if they have a specific breakdown of tasks to be done. Tasks are more efficient if they have a time schedule. And they are more likely to succeed if someone accepts the responsibility.

The project schedule portion of the Orientation & Issues Worksheet does just this. It
- lists the tasks to be done (Action Required)
- shows who is responsible for each task
- sets a schedule for each task

In the simplest form, the tasks (action to be taken) are the six steps of Simplified SNP. Take them right from the Table of Contents. However, it is likely that you will wish to add tasks for problem statement and clarification, for validating the baseline model, running and compiling results...

The vertical-lines section is assigned a calendar set of dates – usually the date shown is "week starting" or "week ending."

Distribution	CSN	By	CN	With	DH	Status of	10/01 Active		
PROJECT SCHEDULE				▼					Notes &
No.	Action Required	Who	9/23-9/28	10/01-10/05	10/08-10/012	10/15-10/19	10/22-10/26		Further Action
1.	Project plan and Schedule	CN/DH							
2.	Problem statement and clarification	DH/DS							
3.	Formulate Variables	CN/KW							
4.	Sensitivity Analysis	KW/BF							
5.	Baseline validation	CN/BF							2
6.	Create Scenarios	CN/DS							
7.	Run and compile scenario results	DS							
8.	Evaluate alternatives	KW/BF							
9.	Present recommendation	CN							
10.	Implementation plan	CN							

Notes: 2. - Meeting of subject matter experts to validate the model results against current network performance indicators.

There is a horizontal line for each task. On that line we show the schedule by a series of brackets as follows:
- bracket at left = earliest start time
- bracket at right = latest completion time
- Horizontal line joining the brackets – duration of that task (Note: progress is posted by filling-in the percent completion within each pair of brackets.)

In our example above, we see that evaluation (line 8) is scheduled to begin on October 18 with two people identified by their initials as KF and BW. This way we get, on one sheet, *what, who* and *when* – the essentials of any delegation or acceptance of an assignment.

At the bottom of the sheet is a space for notes. In our follow through example (repeated here from page 1-2), we can see a note entered as "2". On the form (way to right on line 5) you can see the number "2". This allows us to keep our explanatory notes or comments off the face of our working papers in a disciplined way.

In addition to scheduling, we use the project schedule to record status as planning progresses. A "V" (or arrow head) is placed at the date of the status updating. Then a heavy line within each bracketed period indicates the approximate percent completion for the particular task.

For example, on Tuesday 10/02, the progress is posted (see V at 10/02 and "status of" in upper-right). Task 2 is not yet started and thus is behind schedule.

This illustration begins a follow-through example of Simplified SNP used at Mountain Trail Tonic, Inc. or MTT, (fictitious name), a growing manufacturer and distributor of organic juices. Succeeding chapters will show the key document for each step. In our MTT example, only one facility produces 32-ounce bottles. Demand is increasing and this facility does not have enough capacity to meet expected demand. MTT's network planner has been asked to determine which of six current plants is the best location for a second, 32-oz. "big" bottle manufacturing line. The new capacity will be achieved by upgrading an existing line, but some are easier to upgrade than others. The planner will use an existing sourcing model and software to find the lowest cost manufacturing location. This model includes the Bessemer plant which cannot be upgraded. For convenience and to avoid modifications, Bessemer will be kept in the model but excluded from the alternatives. In addition to cost, the planner must also consider other intangible or "non-cost" factors. All of this information can be seen on the filled-in Orientation & Issues Worksheet.

OUTPUT OF STEP 1

> The output of Step 1 is **THE PROJECT UNDERSTOOD AND SCHEDULED**.

Define th

e Variables

STEP 2: DEFINE THE VARIABLES

Our second step is to define variables, gather data and enter it into our modeling software.

WHAT YOU DO
- Define the model variables.
- Visualize the network.
- Gather inputs: demand, costs, constraints, process parameters.
- List all assumptions.
- Write the formulas to be used.
- Gather, scrub and format the data.
- Formulate the model and enter data to meet model requirements.

Key Document = *Variables Summary Sheet*

Output = *Network model and data ready to run.*

WHY DO YOU DO IT

We cannot get the answers we are seeking unless we select and define our variables correctly. Any mistakes or oversights in this step will become costly later in the project. A variable is any element of the model that can be changed or varied. Thus, "variables" include: design characteristics, parameters, constraints, assumptions, formulas, and even data sources for demand, resources, and costs.

HOW YOU DO IT

1 Complete the heading on the form to identify the worksheet.

2 Define the model design characteristics. Generally these consist of:

1. Locations included
2. Lanes included
3. Products included
4. Product hierarchy
5. Resources included
6. Demand data type
7. Demand data duration
8. Time buckets
9. Unit of measure

3 Make a diagram of the network consistent with (or reflecting) the design characteristics. Use the symbols of Systematic Network Planning for constructing the diagram.

4 Specify parameters for network nodes (locations) and arcs (lanes). Parameter values describe the capabilities and capacities of resources. E.g.: A truck can move 1520 cases; a production line can make 200 cases per hour.

5 Identify the data elements and appropriate sources for demand, resources and cost. Listing data sources helps to validate the model and gain credibility for the modeling process.

6 Define the constraints or limiting conditions that apply to each resource being modeled. For example: A manufacturing line can run only 20 hours a day, 5 days a week for 50 weeks, making its maximum capacity 5000 hours per year.

7 List all assumptions regarding nodes, arcs, resources, parameters and constraints. Examples of such assumptions are: Raw material is available for manufacturing; trucks are always available to move product.

8 Write the necessary formulas for the resources being modeled. For example: Throughput of a resource equals 1/rate of production; Peak capacity equals rate per hour time hours per day times days per week.

9 (Not shown here): Collect, scrub and format the data. Then enter the data into your modeling software and configure to reflect your entries on the Variables Summary Sheet.

VARIABLES SUMMARY SHEET

Company	MTT
Project Name	Big bottle analysis
Sheet	1 of 1
Date	9/23
By	CSN
With	KW, BF
Project Description	32 Oz Bottle line upgrade analysis

MODEL DESIGN CHARACTERISTICS

Locations included	All manufacturing plants, outside bottlers and all distribution centers
Lanes included	All currently active transportation routes between included locations
Products included	All juices sold by MTT
Product hierarchy	Stock Keeping Unit
Resources included	Manufacturing lines, transport trucks, labor, storage equipment at the plants
Demand data type	Historical
Demand data duration	12 Months
Time buckets	Weekly
Unit of measure	Cases

DATA SOURCE SUMMARY

DATA ELEMENTS — DATA SOURCES

DEMAND & RESOURCE RELATED

	DATA ELEMENTS	DATA SOURCES
1.	Item data	ERP Demand Planning
2.	Location data	ERP Demand Planning
3.	SKU data	ERP Demand Planning
4.	Demand data	CIMMS Demand Planning
5.	Sourcing data	CIMMS Demand Planning
6.	Manufacturing lines	Manual
7.	Transportation	Manual
8.	Storage	Manual
9.	Cross handling	Manual
10.		

COST RELATED

	DATA ELEMENTS	DATA SOURCES
1.	Manufacturing costs	SAP
	Variable overhead costs	8111111
	Direct labor costs	8111112
	Changeover costs	8111113
2.	Transportation costs	Steve Hayes (SME)
3.	Cross Handling costs	Logistics P & L
4.	Storage costs	$1.00
5.	Overtime costs	1.5*Direct labor
6.		
7.		

MODEL CONSTRAINTS

1. Fixed capability of manufa... line number 1 can only produce cans).
2. Minimum lots when producing a product.
3. Product costs and throughputs based on demonstrated capability.
4. Maximum weekly run times (80 hours per week at Bessemer).
5. Minimum available weekly run times (140 hours per week).
6. Violation of Min capacity can happen but with a penalty of 1.5*labor costs.
7. Products are produced to cover for a certain duration of time.
8. Use existing lanes to source the products.
9. Infinite transport capacity, which means trucks are available when needed.
10. Storage constraints at the manufacturing plant.

MODEL ASSUMPTIONS

1. Inbound logistics will not be considered.
2. Raw materials available in infinite capacity.
3. Fixed cost of manufacturing will be included in the model.
4. Lot sizing increments will not be considered.
5. A flow through environment for branch, thus storage costs & constraints will be excluded from the model.
6. Cross docking is doubling handling of products without storage.
7. Transport fleet is homogeneous and of infinite capacity.
8. Available capacity of other bottlers will not be included.
9. All rules of contract will be valued and not be violated.

VISUALIZATION

(Diagram showing Outside Bottler, Manufacturing Plants P1–P6, Distribution Centers DC1–DC14, Raw Material Supplier; ANY TO ANY connections)

PARAMETERS

Param. Location	P1 Jonesville	P2 Bessemer	P3 Strasburg	P4 Yuba City	P5 Medina	P6 Bessemer
Lines (count)	6	10	1	1	1	1
Line speeds (cases / hr)	— Demonstrated speeds —					
	120	140	120	120	100	80
Max capacity (hrs per week)	120	120	100	100	100	100
Min capacity (hrs per week)	100	100	100	100	100	80
Storage capacity (pallets)	5000	8000	5000	4000	4000	2500

FORMULAS

1. Transportation cost = X_1*(fleet cost)+X_2*(One way incentive common carrier use)+X_3*(Backhaul factor)+X_4*(Reverse Logistics factor)

   X_1, X_2, X_3, X_4 are trip frequencies

2. Throughput = 1/cases per hour
3. Min Runtimes = Batch size/throughput
4. Peak Capacity = 20*7 = 140 hours
5. Min Capacity = Budgeted capacity/52

SYMBOLOGY

- MANUFACTURING WITH CROSSDOCK
- OUTSIDE BOTTLER
- DISTRIBUTION CENTER
- RAW MATERIAL SUPPLIER
- FINISHED GOODS
- RAW MATERIAL

© Copyright 2007. CHANDRA NATAHAJAN AND RICHARD MUTHER & ASSOCIATES - 775

May be reproduced for in-company use provided original source is not deleted.

... STEP 2 (continued)

SCOPE OR EXTENT OF THE NETWORK MODEL

Every network has design characteristics and these determine the scope or extent of the network model. To get the answers we are seeking, our model must first represent these characteristics:

1. Locations included
2. Lanes included
3. Products included
4. Product hierarchy
5. Resources included
6. Demand data type
7. Demand data duration
8. Time buckets
9. Unit of measure

Locations and lanes included

From the classical theory of network optimization, every problem is based on nodes and arcs. Nodes are the locations in a network. Arcs are the transportation and delivery lanes that join the locations. Before we can solve the network problem, we must define all the relevant locations and lanes. The number of locations and lanes increase the model's complexity, the computer memory required, and the effort to solve the problem. Therefore, include only those locations and lanes for which the model outcomes may be sensitive.

For example, in distribution models with customer ship-to locations, most planners simplify the problem with some sort of aggregation, such as a postal code or zone. When modeling branch warehouse opening and closing, most planners include aggregated customer locations, branch or distribution center locations and possibly manufacturing or supply locations.

In network modeling, lanes "carry" the products or materials between locations. A missing lane leads to demand not being met and mathematical infeasibility for the problem being solved. Therefore the planner must be sure to define all relevant lanes to and from each location. Visualization of the network being modeled helps to avoid missing lanes and the time needed to discover them.

In some models and software, dummy locations may be used as sources and sinks of infinite supply and demand. Dummy locations simplify modeling problems by reducing the number of locations and lanes needed to achieve material flow balance.

As visualized on the Variables Summary Sheet, our MTT model represents multiple raw material suppliers as a single dummy location with one lane to each manufacturing plant. This greatly simplifies the problem. Six outside bottlers are included with MTT's six manufacturing plants. Both act as supplying locations to 42 distribution centers (DCs). Customer ship-to locations are not needed since delivery costs are not affected by the manufacturing plant location of the upgraded 32-oz. bottling line. All transportation lanes between locations are potentially active. Certain products are produced in certain plants and cross docked through other plants. In this way, a branch can receive product from any plant (also shown on the diagram).

Products included

Products or raw material items may be omitted for simplicity when their inclusion is unlikely to impact the model's results. For example, in the MTT model, all raw materials (bottles, lids, packaging) are represented as a single item flow from a dummy location to each plant. Greater detail on raw materials is unnecessary since our model is to find the best assignment of production line output to branch DCs.

If a resource makes or consumes two or more products and changing from one to another affects its capacity, then all the products should be included. But if capacity is unaffected by the presence of multiple products, those not relevant to the outcome should be excluded to simplify the problem and save time.

Product hierarchy

Product hierarchy refers to the ways in which data about items or stock-keeping units may be grouped, rolled-up or aggregated. The purpose of aggregation is to avoid unnecessary effort while assuring sufficient detail and accuracy. Usually, short-range operational planning models need higher accuracy and therefore more granularity and detail about specific items. In contrast, long-range strategic planning is based on approximate forecasts. Therefore network modeling can be adequately performed at a higher level of aggregation, saving time and reducing effort.

Planners often group items of a common type or characteristic such as size, material, or package type. In a beverage company, modeling might be aggregated by package family, such as 16oz. or 32oz., or by brand, or product style. Our MTT example includes all juice products, both purchased and manufactured.

In transportation models, if the individual items being carried do not matter, the planner may use transport units such as: inner pack, master pack, case, pallet or container.

Grouping by velocity – fast movers, medium, and slow – may simplify warehouse capacity models when the space or resources devoted to specific items, product families or packages do not matter. Grouping seasonal and non-seasonal products may also simplify capacity planning models for production and warehousing.

Resources included

In network modeling, locations and lanes have resources which, in turn, have costs and capacities. Typical location resources are: manufacturing lines, hours of availability, labor hours, storage positions, number of docks... Typical lane resources are trucks and delivery drivers. Which resources to include in the network model will be determined by which costs and capacities are relevant to the planning objectives.

When they are not assumed to be infinite, resource capacities must be specified in appropriate units of measure. Constraints set upper and lower limits on resource capacities. Parameters specify the number of relevant resources and their constraints. In our MTT model, the resources are: manufacturing (bottling) lines, transport trucks, labor, and storage equipment at the plants. Constraints and parameters for these resources are: the number of bottling lines at each location and their speeds, their maximum and minimum capacities in hours available per week, and storage capacities in pallets. Note that transportation is a resource, but it will be considered as infinitely available.

Demand data type (historical or forecast)

Demand data refers to orders, sales and/or shipments. Such data can be historical or forecast. Model validation always requires historical data. But when modeling alternative scenarios, we must decide whether to use historical or forecast data. In some cases, history may be preferred to forecasts. For example, when planning for new products, it may be wise to use historical data for a similar existing product rather than a forecast that may be subject to large error. On short-term projects, when the near-term future may be like the recent past, using historical data may be quick and convenient and less subject to speculation. In other cases, history may bear no resemblance to the expected future and would give misleading results. In our MTT example, the planner is seeking to cost justify a production line upgrade using only historical data. The model will show much money would have been saved had a second line been upgraded last year.

Demand data durationco

When demand varies greatly from period to period, we must decide which periods to include. In the figure at right, demand in the first half of the year is higher than the second. So if the planning is for the peak periods, only the first half need be used. But if planning for high and low periods, the full year should be used. As a rule, include enough periods to capture the demand variation that is relevant to the planning situation.

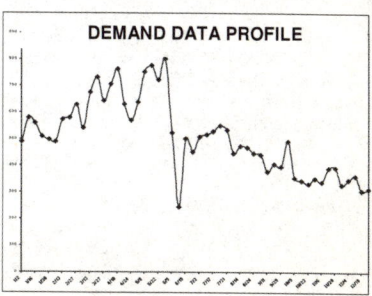

... STEP 2 (continued)

When using forecast data, be sure to include enough future periods to cover the useful economic life of the principal assets being planned. If our plan is for seasonal inventory build, then we only need a forecast for the upcoming season. But if our plan is to locate a production line with a 10-year useful life, then we should include enough periods to cover much if not all of the line's useful life.

Time buckets

Time buckets aggregate demand periods. Weekly or monthly aggregations show less variability than daily. For short term operating plans use smaller time buckets so that relevant variability is considered. For strategic and long range planning, larger time buckets are fine. When we are unsure, or when our model will examine both operational and strategic choices, we should extract data in the smallest available time period and aggregate later to larger buckets.

Unit of measure

Unit of measure refers to the demand data unit and to the level of aggregation when presenting model results. For example, cost per case or cost per pallet. For short term and operational planning, use smaller units of measure and less aggregation. Organizational metrics also play a role. If cost per case is important to management then this unit of measure may be best. Choose the unit of measure carefully and stick with it to avoid tedious rework, conversions, errors and confusion.

Our MTT model will use 12 months of historical demand data in weekly buckets, measured in cases.

VISUALIZATION

Visualizing the network helps to confirm scope and avoid oversights such as missing locations or lanes. SNP uses an adaptation of industry-standard operation process charting symbols. But any well-labeled symbols can be used.

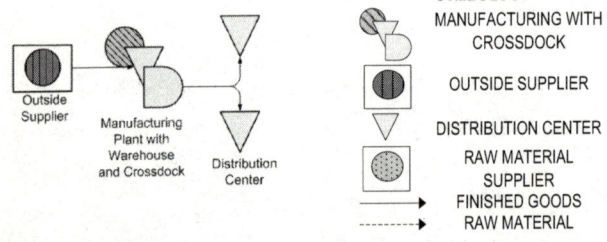

PARAMETERS AND THEIR VALUES

Parameters and their values describe the numbers, capabilities and capacities of resources. These are summarized in a table on the Variables Summary Sheet. When resources are constrained by physical limitation or by operating policies, we use parameter values to express these constraints. In our MTT example, the parameters include number of lines, maximum and minimum capacities expressed in maximum and minimum hours of operation, pallets of storage capacity, and line speeds. For example, Plant 1 (P1) has six manufacturing lines and their line speeds are their demonstrated speeds, meaning that the modeler will use the line speeds in cases per hour normally used by the production planners when scheduling each line.

USE OF APPROPRIATE DATA SOURCES

Often, data for the same model variables may be obtained from two or more sources. Data values may differ based on the way in which the data were collected or reported. Generally, we should favor the data sources that are used to prepare the company's official financial reports, since model results will typically be validated by comparison to reported results. However, there may be times when other data sets, perhaps maintained locally at a plant or distribution center, will provide more accurate results. If these data will be used, be ready to explain why they are being used in place of "official" data and sources. Identify all data sources on the Variables Summary Sheet. This helps to build credibility for the model, prevent costly rework, and assure acceptance of model results.

In our MTT example, data about products will come from the CIMMS Demand Planning information system. Data about resources will be manually entered. Manufacturing cost data will come from transaction records in the SAP ERP system. Other cost data will come from various sources.

Demand & Resource Related Data Elements

All models include basic data elements about demand and resources. Some elements are generic to any model. Others may be particular to the problem and project. Common elements include:

Demand-related:
1. Items
2. Locations
3. SKUs
4. Sources
5. Customers

Resource-related:
1. Manufacturing equipment and labor
2. Transportation equipment
3. Storage locations
4. Material handling equipment and labor
5. Delivery equipment

Cost-related Data Elements:

Cost data elements are always needed since the purpose of network models is to optimize for low cost or maximum profit. Common elements include:

1. Raw material or purchase costs
2. Direct and indirect labor
3. Changeovers and set up cost
4. Transportation and/or delivery cost
5. Handling costs
6. Storage costs
7. Overtime costs
8. Cost of capital
9. Inventory carrying costs
10. Fixed cost of production line / real estate

SCRUBBING AND FORMATTING THE DATA

Data is never all clean and ready to use. This is a big problem in many modeling efforts. Common problems include: invalid item numbers, presence of discontinued items, omission of new items, missing item descriptions or other null fields that should contain valid entries. These and other problems may stem from data entry errors or from the data selection procedure. The planner should always "sanity check" totals for: numbers of items, total shipments or demand, maximum shipment quantities, negative values, especially for inventory, and the like. Also check to be sure that entry of the data into the model has not itself resulted in errors, such as improper assignment of items to locations, or other invalid associations. It is essential to perform these checks before running the model. Otherwise it will be difficult to isolate data issues from modeling issues during validation, and you will significantly delay your project. Even when checking in advance, allow plenty of time for scrubbing and formatting the data.

CONSTRAINTS

Resource constraints specify limitations on the various resources being modeled. Resource constraints may stem from physical limitations in the actual network or from operating policies and practices. A policy constraint might state that manufacturing lines cannot run more than 140 hours per week and must run at least 80 hours. Clearly, up to 168 hours are available, and we could run a line for less than 80 hours.

In addition to resource constraints, we must also define modeling constraints. These define required behavior within the model. For example: All demand must be satisfied. Such a model constraint may lead to infeasibility if satisfying this constraint would exceed our resource constraints. Resolving such conflicts and infeasibilities will be worked out in Steps 3 and 4.

...STEP 2 (continued)

A systematic approach to defining, listing and building constraints is much faster than random identification. An effective procedure is to walk through the visualization and asking what are the limiting conditions for each location and lane, listing them as we go. Examples of typical network modeling constraints include:

1. Maximum capacity (run time) of production limited to 120 hours per week
2. Lot-size constraints e.g. run at least 1000 cases each time the product is produced.
3. A given production line can only run certain products.
4. Transport capacity on a lane is infinite (unconstrained); or is limited to 2000 cases per day.
5. A branch warehouse can be supplied from no more than two producing locations.
6. Shelf life constraints, e.g. products can be stored for no more than "X" days.
7. Inventory constraints, e.g. production runs should yield at least 5 days of supply.
8. Storage capacity, e.g. a branch cannot store more than 5000 pallets.
9. Shipment size constraints, e.g. a branch cannot send less than a full truck load to any other branch.
10. Facility capacity, e.g. a dock door cannot handle more than 40 truck loads per day.
11. Percentage customer service levels. Service level is 99% which means that the customer demand can be met 99% of the time.

Note that some constraints are "hard" and some are "soft." By hard, we mean that they must be observed. By soft, we mean that the constraint may be violated at a specified penalty cost. When in doubt, use soft constraints since these will result in fewer infeasibilities and less rework. In the examples of above, storage capacity would typically be a hard constraint. Maximum capacity (run time per week) of 120 hours might be set as a soft constraint with an overtime penalty cost incurred for running up to 140 hours, or even to the full 168 hours available.

ASSUMPTIONS

Assumptions are statements accepted by the modelers as true without proof. Assumptions are used to clarify model scope, to simplify the model, and to clarify the manner in which some variables will be treated. Listing our assumptions helps our subject-matter experts and our decision-makers understand our model and its results. It also makes them easier to challenge and quicker to change when needed. Examples of MTT's assumptions and their modeling implications are listed below:

ASSUMPTIONS	IMPLICATIONS
Inbound logistics costs will not affect the choice of line location.	Transport costs from raw materials suppliers to plants will not be considered. Any backhaul opportunities will be computed as a part of the transportation cost formula for each lane.
Raw material is available in infinite supply	Raw material is available when it is needed for production
Transportation will be available in infinite supply	Some combination of private fleet and common carrier trucks will always be available on a particular lane to move the product.

FORMULAS

Formulas may be used to calculate costs or resource performance not available directly as source data. Formulas should be written down and included on the Variables Summary Sheet. This helps to explain the model to others and to confirm the planned approach. Our MTT example uses the following formulas. These are typical for network models.

1. Throughput = 1/cases per hour, where cases/hour is the demonstrated line speed. Thus, if line speed is 200 cases per hour, then throughput = 1/200 or 0.005 hours per case.

2. Minimum Run Time = Batch Size / Throughput. This expression assumes that each item is produced in a minimum batch or lot size. Dividing this quantity (in cases) by throughput (1/cases per hour) gives the minimum time for this item on the production line.

3. Peak Capacity = 20*7 = 140 hours. This formula represents capacity as hours of run time available per week, where 20 is hours of run time available each day and 7 is the number of days in a week.

4. Minimum Capacity = Budgeted Capacity / 52. This formula also represents capacity as weekly hours of run time with a minimum number of hours set by budgetary policy. For example, if a line must run at least 5400 hours in a year – roughly 2 shifts – then this formula says that each line must run at least 5400 / 52 weeks in a year, or 104 hours per week.

5. Transportation cost is defined as:

   Transportation cost = X_1*(2-way private fleet cost)

   +X_2*(One way incentive for common carrier use)

   +X_3*(Backhaul factor) +X_4*(Reverse logistics factor, i.e. container and damaged goods return),

   where X_1, X_2, X_3, X_4 are trip frequencies for each lane.

This formula shows that the planner is using a weighted average approach to estimate a single transportation cost for each lane, rather than modeling each kind of transportation as a separate resource on each lane. The latter approach would significantly increase the complexity of the model without significantly increasing the precision of the results

MODELING SOFTWARE

Logistical networks are modeled using specialized software. A number of commercial products are available, ranging from simple and inexpensive to complex and costly. Some stand alone as purely analytical tools. Others are bundled with Enterprise Resources Planning (ERP) software and configured specifically for logistical network modeling. Most such software uses the CPLEX optimization software engine for mixed-integer linear programming. This engine optimizes – meaning it seeks to maximize or minimize an objective function subject to constraints. The objective function is typically defined to minimize costs of interest, or to maximize profits. Most modeling software has a three-part structure:

- User interface to enter and view data and results
- Mid-layer where mathematical equations are formulated
- Optimization engine which solves the problem as formulated

A more detailed discussion of modeling software is beyond the scope of this booklet, and our Systematic Network Planning procedure is not dependent upon any specific algorithm or software product.

Note that modeling software does not typically consider the intangible factors or consequences of network planning decisions. Factors such as risk or ease of implementation, or flexibility are outside the scope of modeling software and must be addressed by a larger and more comprehensive planning procedure such as Simplified SNP.

OUTPUT OF STEP 2

The output of Step 2 is **A MODEL AND DATA READY FOR VALIDATION.**

Analyze the

sensitivities

STEP 3: ANALYZE THE SENSITIVITIES

Next, we run our model, troubleshoot for infeasible results and fine-tune until our results match actual network performance.

WHAT YOU DO

- Run the model.
- Identify infeasibilities, troubleshoot and create a model free from infeasibilities.
- Run again to replicate actual performance of the current network and measure the variance.
- Adjust the model until variance from actual performance is acceptable.
- Summarize model results and actual network statistics.

Key Document = *Validation Worksheet*

Output = *Validated baseline model that replicates current performance.*

WHY YOU DO IT

To have credibility, our network model must replicate the actual performance of the current network. Without a validated baseline model, we cannot trust model results for future scenarios and alternative network plans. During the validation procedure, we will also learn the sensitivities of demand and cost variables to changes in resource parameters, constraints, and assumptions.

HOW YOU DO IT

1. Complete the heading on the form to identify the worksheet.

2. List the variables to be validated and their units of measure. If not already specified in Step 2, use units of measure that are familiar to the organization and used to measure current network and facility performance.

3. Run the model and post the result for each demand and cost variable.

4. Obtain and post the actual value for each variable – in the current network. Typically these will be available from accounting and financial information systems, or from an Enterprise Resource Planning (ERP) system. Productivity reports and "scorecards" may also be sources of actual performance.

5. Compute the variance (difference) between model results and actuals.

6. Confirm that variances are within acceptable limits – typically a small percentage from actual.

7. Use notes to explain differences that are significant or lessons learned for future modeling.

8. Check to be sure that model constraints are observed. If relaxed or modified from Step 2, enter the revised constraint and explain with a note. Also enter any new constraints that become necessary.

9. Enter notes to explain the variances and any constraints not honored.

BASELINE VALIDATION WORKSHEET

Company	MTT	Project Name	Big Bottle Analysis
By	CSN	With	KW, BF
Date	9/23	Sheet	1 of 1

VARIABLES

	VARIABLES	UOM	MODEL RESULTS	ACTUALS	VARIANCE	OK	NOTES
DEMAND							
1.	Outflow at Sommersville production center	Cases	10,985,632	10,916,130	69,502	✓	1
2.	Outflow at Jonesville production center	Cases	7,845,978	7,880,862	(34,884)	✓	2
3.	Outflow at Briansville production center	Cases	1,288,448	1,318,526	(30,078)	✓	3
4.	Outflow at Vicksburg production center	Cases	1,071,289	1,085,172	(13,883)	✓	4
5.	Outflow at Madison production center	Cases	1,405,731	1,442,651	(36,920)	✓	5
6.	Storage at Sommersville production center	Pallets	7,146	7,265	(119)	✓	6
7.	Storage at Jonesville production center	Pallets	4,754	4,855	(101)	✓	6
8.	Storage at Briansville production center	Pallets	3,849	3,953	(104)	✓	6
9.	Storage at Vicksburg production center	Pallets	4,319	4,224	95	✓	6
10.	Storage at Madison production center	Pallets	3,689	3,626	63	✓	6
COST							
1.	Manufacturing cost at Sommersville production center	Cost / Cs	$1.0820	$1.0806	$0.0014	✓	
2.	Manufacturing cost at Jonesville production center	Cost / Cs	$1.0838	$1.1087	($0.0249)	✓	7
3.	Manufacturing cost at Briansville production center	Cost / Cs	$1.0801	$1.0960	($0.0159)	✓	
4.	Manufacturing cost at Vicksburg production center	Cost / Cs	$1.0873	$1.1144	($0.0272)	✓	7
5.	Manufacturing cost at Madison production center	Cost / Cs	$1.0805	$1.0804	$0.0001	✓	
6.	Transportation cost	Cost / Cs	$1.7309	$1.7256	$0.0053	✓	
7.	Cross docking cost	Cost / Cs	$0.0500	$0.0510	($0.0009)	✓	8
8.	Purchasing cost (cost for finished goods from outside bottlers)	Cost / Cs	$6.1600	$6.1600	$0.0000	✓	
9.							
10.							

CONSTRAINTS

		OK	
1.	Fixed capability of manufacturing. (line number 1 can only produce cans).	✓	
2.	Minimum lots when producing a product.	✓	
3.	Product costs and throughputs based on demonstrated capability.	✓	
4.	Minimum weekly run times (80 hours per week at bessemer) .	✓	
5.	Maximum available weekly run times (140 hours per week).	✓	
6.	Violation of Min capacity can happen but with a penalty of 1.5*labor costs.	✓	
7.	Products are produced to cover for a certain duration of time.	✓	
8.	Use existing lanes to source the products.	✓	
9.	Infinite transport capacity; which means trucks are available when needed.	✓	
10.	Storage constraints at the manufacturing plant.	✓	

NOTES / EXPLANATION:

1.	Sommersville produced 70,000 fewer cases due to a one-week production line shut down.
2.	Jonesville produced cases for Sommersville during the line shut down.
3.	Briansville produced cases for Sommersville during the line shut down.
4.	Vicksburg produced 5,000 cases for Sommersville. Remaining variance results from 2-day marketing promotion.
5.	Madison had a promotional campaign that was not included in the model.
6.	The variances in stored pallets are due to lot sizing assumptions.
7.	The variances of 2 to 3 cents per case are due to purchase price variances on raw materials.
8.	Cross docking cost was higher due to trans-shipping and double handling during the line shut down at Sommersville.
9.	
10.	

© Copyright 2007. CHANDRA NATARAJAN AND RICHARD MUTHER & ASSOCIATES - 777   May be reproduced for in-company use provided original source is not deleted.

... STEP 3 (continued)

TROUBLESHOOTING INFEASIBILITIES

Typically when a model is first run and its results reviewed, one or more constraints will have been violated. Such instances are called "infeasibilities" and they must be removed. Common examples are: exceeding or under using available capacity; lot size too high or low; operating hours too high or low; inventory above a maximum or below a minimum... Common causes include: missing lanes, locations or items; improperly defined constraints; supply and demand imbalance, problems with data... A practical procedure for troubleshooting and removing infeasibilities is:

- Relax some of the "hard" constraints by making them "soft" policy constraints. These can be violated but incur a high penalty cost when doing so.
- Run the model again and identify the location or lane with the highest cumulative penalty cost.
- Examine each variable, constraint and assumption that might be contributing to the penalty cost.
- Once contributing factors have been identified, make appropriate adjustments, such as:
  - tweak (modify slightly) the appropriate resource parameter,
  - add lanes,
  - add locations
  - redefine constraints
  - redefine assumptions
- Reset the constraints in question to "hard" and re-run the model.
- Review the results for infeasibility, and repeat until all constraints are honored.

Another useful practice is to populate the initial model run with data for a single, universal product – one that is found at every location and moves on every lane. Then get the results for this universal product to match its actuals before populating the model with additional or all remaining products.

Troubleshooting is an iterative process. The number of iterations is significantly reduced by following Steps 1 and 2 of Simplified SNP but the process may still take several days.

FINE-TUNING

Once all constraints are honored, the planner runs and fine-tunes the model to establish a baseline that replicates the actual performance of the current network. Fine-tuning typically involves adding resources, adding or re-defining constraints and formulas, and tweaking parameter values until model results match actual results for a chosen historical period. For example: If the number of cases produced in the model matches actual cases produced, but the production hours used exceed the actual, the planner may adjust the production line rate parameter to get the hours closer to actual.

Fine-tuning may take several hours to several days, depending upon the scope and complexity of the model. The network planners should decide with management how much variance is acceptable, recognizing that no model can fully capture all operational reality. The smaller the acceptable variance, the more time will be required for fine-tuning.

STAYING FOCUSED ON KEY VARIABLES

Logistical modeling software may produce an overwhelming number of statistics and much time could be spent examining them. You will save time by focusing and reporting first on those variables that you defined in Step 2, and by limiting additional review to variables that the organization currently measures. Other statistics from the software may be helpful for troubleshooting but may distract or confuse your audience if reported on the Baseline Validation Worksheet.

If the model does not produce statistics in the desired unit of measure, be sure to transform them to the units that your audience expects. For example the software might produce results in units when pallets are the normal way of understanding volumes. Transforming to an expected unit will improve understanding and speed up validation.

COMMON VARIABLES FOR SENSITIVITY ANALYSIS

While the specific variables to be analyzed are unique to each project, the ones below are common.

Demand-related	Cost-related
•Resource utilization (production lines, storage facilities, trucks, labor)	•Manufacturing (labor + variable overhead)
•Demand met	•DC replenishment (fuel cost + other trucking cost per mile, including labor)
•Demand unmet	•Storage equipment
•Pallets / Cases / Units sold	•Warehousing (labor +variable overhead)
•Units produced	•Raw material (including transportation)
•Total hours of production/Operation	•Purchasing (activity costs for the purchasing department)
•Service times	•Service (cost of taking an order or merchandising in a store)
•Total overtime	•Distribution (delivering to a retail store or customer location)
	•Fixed costs (occupancy, depreciation)

In our MTT carry-through example, the demand-related variables of interest are the outflows (production) and storage at each plant. (See Page 4-2). The cost-related variables are: manufacturing cost at each plant, transportation, cross-docking, and purchasing. Actual performance for the current network was obtained from the company's ERP system. Per the model's design characteristics in Step 2, 12 months of historical production data was obtained for each plant (production center) and compared to the model's results.

Model results for outflow cases at Sommersville were 10,985,632 compared to 10,916,130 actually produced, for a variance of 69,502. This variance is attributed in Note 1 to a one week production line shut down in Sommersville.

The model's cross docking cost of $0.05/case compares to actual costs of $0.051. In Note 8, this variance is attributed to the same Sommersville line shut down, during which time its demand was filled by cross-docking with associated double handling of product.

Actual manufacturing cost at Jonesville was $1.1087 per case while model results were $1.0838, for a variance of $0.0249 per case. This was found to be caused by a purchasing price variance as explained in Note 7.

All constraints from Step 2 are observed and none needed to be changed.

OUTPUT OF STEP 3

> The output of Step 3 is a **VALIDATED BASELINE MODEL**

Create

scenarios

STEP 4: CREATE SCENARIOS

In this step we develop and model scenarios representing alternative network plans.

WHAT YOU DO

- Vary the baseline model to reflect the choices and decisions at hand.
- Group model changes into alternative scenarios, each reflecting a different network plan.
- Define optimistic, pessimistic and most likely cases for each scenario.
- Run the model for each case of each scenario.
- Document the results of each model run.

Key Document = *Scenario Summary Sheet*

Output = *Model results for alternative network plans*

WHY YOU DO IT

There are always alternative ways to configure a logistical network – so many that planners and their managers can become overwhelmed by the choices. An orderly and comprehensive approach assures that all promising plans are considered and that each is tested for best-case, most-likely, and worst-case operating situations.

HOW YOU DO IT

1 Complete the heading on the form to identify the worksheet.

2 Assign a Roman numeral and give a descriptive name to each scenario. Then for each:

3 Record the changes needed to the baseline model set-up:

- Demand-related: Add and/or drop products and/or locations; change allocations of demand to locations.
- Resource-related: Add and/or remove resources; change parameters on existing resources.
- Lane-related: Add and/or drop lanes; change characteristics on existing lanes.

It may also be useful to note where the baseline set-up remains the same.

4 Record additional constraints and/or changes to the baseline model, and confirm those still in force.

5 Record additional assumptions and/or changes to the baseline model, and confirm those assumptions still valid.

6 Visualize the scenario's network with a map or diagram that shows its differences from the baseline.

7 Define three cases for the scenario (identified by lower-case letter), listing variations to demand, resources, lanes, constraints and assumptions:

   a. An optimistic or best case in which demand, resource parameters, constraints, and assumptions are set to their most desirable states.
   b. A most-likely case in which demand, resource parameters, constraints, and assumptions are set in between their most and least desirable states.
   c. A pessimistic or worst case using least desirable states.

8 Create an instance or version of the model for each scenario case, run the models and record the resulting costs.

9 Briefly summarize other results of the scenario model runs in terms of demand, resources and flow.

10 Record any additional notes that help to understand the scenario or explain its results.

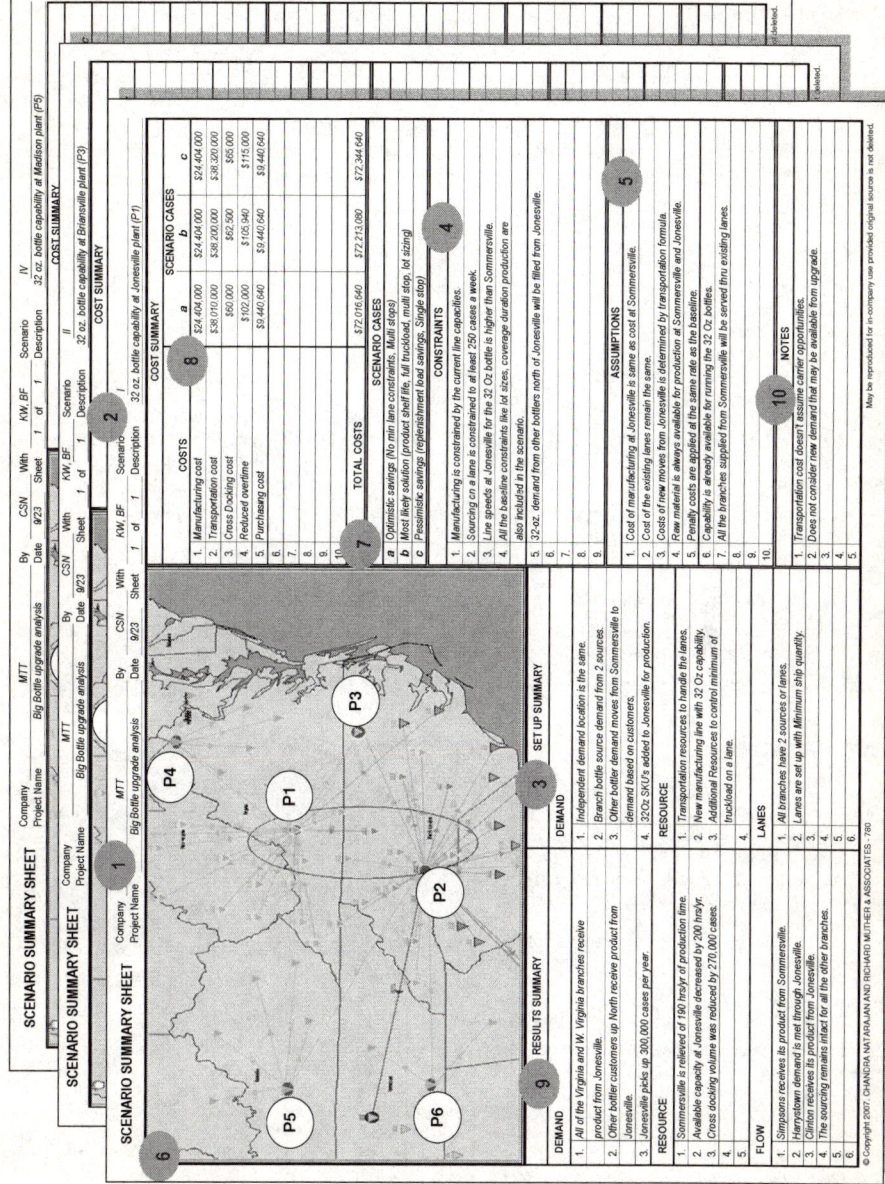

... STEP 4 (continued)

SCENARIO CHANGES

The purpose of a scenario is to investigate an alternative network design. Alternatives represent different ways that the network may be designed or managed in response to expected demand and operating conditions. There are a myriad of scenarios for any planning situation, so the planner must be insightful to avoid getting bogged down. Typically between three and five scenarios are sufficient to explore the most promising network plans. Each should represent a significantly different but feasible plan.

After conceptualizing scenarios, planners must next identify the critical handful of variables that need to be changed from the baseline model in order to properly represent each alternative. Typical changes include:

1. Addition and/or deletion of resources
2. Changes to the characteristics (parameter values or costs) of existing resources
3. Addition and/or deletion of new sourcing lanes
4. Changing characteristics or capacities of sourcing lanes
5. Changes to the values of constraints, additional constraints or removal of constraints
6. Additional demand, including:
   - New products or items
   - New locations
   - New or re-allocations of demand among locations\
   - Removal of demand via removal of products, items, locations, or allocations.

In our Mountain Trail Tonic carry-through example, four scenarios each place the upgraded 32-ounce bottling line at a different plant. In Scenario I (32-oz. capability at Jonesville) summarized on Page 4-2, additional changes from the baseline model include:

Demand – is still the same since this project uses 12 months of demand history. But in this scenario, branches can receive products from either the Jonesville or Sommersville plant.

Resource – additional transportation capacity (trucking) as needed.

Lanes – all branches have two source plants, as stated above.

Constraints/Parameters – shipments to branches must be full truckloads; the upgraded line at Jonesville will be faster than the existing at Sommersville; 32-oz. demand from other bottlers north of Jonesville will be filled from there.

Assumptions – cost of new moves from Jonesville will be determined by the baseline transportation formula.

NETWORK DIAGRAM

It is important to visualize each scenario's network with a map or diagram. This is especially the case when a scenario changes the network structure (lanes and/or locations) from the baseline model as visualized on the Variables Summary Sheet from Step 2. In MTT's Scenario I, the baseline scenario lane between Jonesville and Sommersville is removed and some branches are reassigned to different plants.

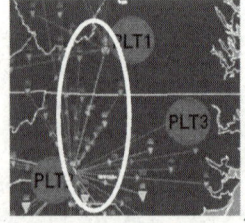
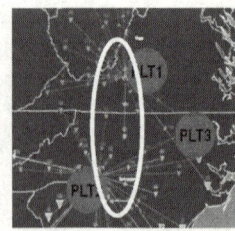

Baseline Network   Scenario I Network

SCENARIO CASES

Often, those managing the network and its resources will reject the projected cost savings from an initial scenario run as too optimistic even though the model was validated in Step 3. In response, the planners and "subject matter experts" may have to adjust assumptions, parameters or constraints until the scenario run yields a more modest improvement – one that the line organization is willing to be accountable for obtaining. But such a "pessimistic" outcome may be resisted by the planners who rightfully have faith in their validated model. The SNP procedure anticipates this give and take, and allows for optimistic, pessimistic and most-likely cases. These can then be presented to management to show the likely range of scenario outcomes.

SCENARIO RESULTS SUMMARY & COSTS

Each scenario case is created and run and its infeasibilities fixed. It is then fine tuned and its results summarized in terms of demand, resources, flows and cost.

In our MTT example, when the most-likely case for Scenario I was run, the optimizer moved 300,000 cases of 32-oz. production from Sommersville to Jonesville. Of these, 270,000 were formerly cross-docked through Jonesville to its branch DCs. In terms of resources, Sommersville was relieved of 190 hours of production, while 200 hours were added to Jonesville (thus decreasing its available capacity). The 100-hour difference is due to the higher speed of the upgraded line in Jonesville. In terms of flow, Simpsons (DC 23) could not satisfy the full truckload constraint from Jonesville and was reassigned to Sommersville. DCs 11 & 12 – Harrystown and Clinton – along with all the Virginia and West Virginia DC's, received all of their 32-oz. products from Jonesville.

In terms of total costs, there is virtually no difference between the optimistic most-likely cases and pessimistic cases. This is because the manufacturing and purchasing costs are the same in all cases and the differences in transportation, cross-docking and overtime are insignificant relative to manufacturing and purchasing.

Still, there is a $190,000 transportation cost difference between Case a. and Case b. and $310,000 between Case a. and Case c. These are due to differences in replenishment movements between plants and branches. In the optimistic and most likely cases, replenishment trucks were routed to multiple stops at more than one DC while in the pessimistic case, each branch DC receives its own dedicated, single-stop delivery.

RESULTS SUMMARY	
DEMAND	
1.	All of the Virginia and W. Virginia branches receive product from Jonesville.
2.	Other bottler customers up North receive product from Jonesville.
3.	Jonesville picks up 300,000 cases per year.
RESOURCE	
1.	Sommersville is relieved of 190 hrs/yr of production time.
2.	Available capacity at Jonesville decreased by 200 hrs/yr.
3.	Cross docking volume was reduced by 270,000 cases.
4.	
5.	
FLOW	
1.	Simpsons receives its product from Sommersville.
2.	Harrystown demand is met through Jonesville.
3.	Clinton receives its product from Jonesville.
4.	The sourcing remains intact for all the other branches.
5.	
6.	

COST SUMMARY				
	COSTS	SCENARIO CASES		
		a	b	c
1.	Manufacturing cost	$24,404,000	$24,404,000	$24,404,000
2.	Transportation cost	$38,010,000	$38,200,000	$38,320,000
3.	Cross Docking cost	$60,000	$62,500	$65,000
4.	Reduced overtime	$102,000	$105,940	$115,000
5.	Purchasing cost	$9,440,640	$9,440,640	$9,440,640
6.				
7.				
8.				
9.				
10.				
	TOTAL COSTS	$72,016,640	$72,213,080	$72,344,640
SCENARIO CASES				
a	Optimistic savings (No min lane constraints, Multi stops)			
b	Most likely solution (product shelf life, full truckload, multi stop, lot sizing)			
c	Pessimistic savings (replenishment load savings, Single stop)			

When large costs such as manufacturing and purchasing here are identical in each scenario case, and in all scenarios, their inclusion will reduce the apparent and percentage differences in total costs. For this reason, it may be wise to remove such costs from the summary and compare only those that are subject to change. But, if the planner's audience wonders about manufacturing and purchasing cost, their inclusion will confirm that they are not affected.

Often, some cost categories of concern at the outset of modeling prove to be insignificant once results are in. Or, an important stakeholder may have a fixation on some minor element of network cost. Notice that for MTT, cross-docking and overtime are insignificant, as well as their percentage changes. In this example, the planner has included these costs to satisfy the concerns and interests of key decision-makers. Such findings help to understand which costs will drive the evaluation of alternatives in Step 5 and which costs can be safely ignored.

NOTES/EXPLANATIONS

When special or detailed aspects of a scenario do not fit the pre-defined categories of the Scenario Summary Sheet, the Notes/Explanations section is a good place to capture them. In our MTT example, the reference to "carrier opportunities" notes that increased use of potentially-cheaper common carriers instead of the baseline model's private fleet has not been considered. During the modeling, it may be impractical to get rates from a transportation department or from carriers directly for every scenario and case. Such a note reminds the planner to revisit this opportunity later should this scenario's plan be selected or among the top contenders.

OUTPUT OF STEP 4

The output of Step 4 is **MODEL RESULTS FOR ALTERNATIVE NETWORK PLANS.**

Evaluate the

Alternatives

STEP 5: EVALUATE ALTERNATIVES

Here we evaluate costs and intangibles factors and select the best network plan.

WHAT YOU DO

- Identify and name each alternative network plan.
- Compare annualized costs and identify the plan with the lowest total costs.
- Rate and score the performance of each plan against relevant intangible factors.
- Compare the performance of plans on cost and intangibles.
- Agree on the best plan considering both costs and intangibles.

Key Document = *Alternatives Analysis Worksheet*

Output = *Alternative network plans evaluated and the best plan selected.*

WHY YOU DO IT

The lowest cost plan may not be the best when important intangible factors are considered. Only by making an evaluation of the costs *and* the intangible factors can you objectively and impartially choose the best network plan.

HOW YOU DO IT

1 Complete the heading on the form to identify the worksheet.

2 Post the summarized costs from each Scenario Summary Sheet, omitting any that are not relevant to the evaluation, e.g. costs that are the same for all scenarios or costs that are insignificant to the outcome.

Note: Where capital investment or one-time expenses will be incurred to make network changes, divide the total of these expenses by the useful life of the changes, in years. This will give you an annualized look at investment cost, in addition to the operational costs modeled in Step 4. While not a financial *justification*, this simple comparison will often identify the most economical alternative. You must still cost justifiy any investment by working with your accounting or finance department.

3 Enter the corresponding costs for the baseline model.

4 Total the costs for each alternative and compute the difference of each from the baseline.

5 List all factors, considerations or objectives the company wants the network plan to achieve. Use care in the wording, so the factors read positive, for example, "Freedom from breakdown", not "Breakdowns". (For examples of intangible factors, please see page 5-4.)

6 Ask your approvers to select the most important factor and assign to it weight of 10, Then ask them to weigh the importance of each other factor relative to the most important (10). Indicate each selected weight on the form, and record by whom the weight values were determined (in box above).

7 Ask your network operations managers to rate for each factor, the effectiveness of each alternative in achieving that factor's objective. Use A, E, I, O and U to represent the descending order of effectiveness. Enter these letters in the small rectangular boxes on the form. And record the name(s) of the person(s) doing the rating.

8 After rating all alternatives for each factor, convert letters to numbers (A=4, E=3, I=2, O=1, U=0) and multiply by the respective weight values. Enter the resulting weighted-rated values on the form.

9 Down-total the weighted-rated values for each alternative plan; enter on the form; and record by whom the tally was made. The plan with the highest total should be the "winner" -- subject to cost evaluation above.

10 Record any explanatory notes.

ALTERNATIVES ANALYSIS WORKSHEET

Company	MTT	Project Name	Big Bottle Analysis
By	CSN	With	BF, KW, CN
Date	10/21	Sheet	1 of 1

COST ANALYSIS

By: *CSN* With: *BF, KW*

COST SUMMARY	BASELINE	ALTERNATIVES I	II	III	IV
1. Manufacturing Costs (Std Costs)	$24,404,000	$24,404,000	$24,404,000	$24,404,000	$24,404,000
2. Transportation Costs	$38,850,000	$38,200,000	$38,315,000	$38,340,000	$38,500,000
3. Cross Docking	$76,000	$62,500	$65,200	$67,500	$66,500
4. Reduced Overtime (Manufacturing variance)	$143,000	$105,940	$107,900	$112,200	$111,000
5. Purchasing Costs (omitted since equal for all)	NA	NA	NA	NA	NA
6.					
7.					
8.					
9.					
10.					
TOTAL ANNUALIZED COSTS	$63,473,000	$62,772,440	$62,892,100	$62,923,700	$63,081,500
ANNUALIZED SAVINGS OVER BASELINE		-$700,560	-$580,900	-$549,300	-$391,500

INTANGIBLE ANALYSIS

Weights by: *KW, BF, CN*   Ratings by: *CN, TT, RH*   Tally by: *CSN*

FACTOR / CONSIDERATION	WT.	BASELINE	ALTERNATIVES I	II	III	IV
1. Capacity relief	10		O 10	A 40	E 30	O 10
2. Ease of implementation	6		A 24	E 18	O 6	O 6
3. Union environment	5		U 0	U 0	U 0	I 10
4. Organizational structure	4		I 8	I 8	U 0	I 8
5. Condition of the line	1		U 0	I 2	I 2	O 1
6. Availability (less risk of line shutdown)	8		X	I 16	I 16	I 16
7.						
8.						
9.						
10.						
TOTAL INTANGIBLES EFFECT			42	84	54	51

NOTES

1. Capacity savings is due to the penalty of overtimes.
2. Briansville crossdock only limited products due to space constraints
3. Madison is the only unionized plant.
4. Risk of line shut down is a negative impact condition. High risk is given lower rating.

SCENARIO NAME

BASELINE	Current network performance
I	32 OZ Bottle capability at Jonesville
II	32 OZ Bottle capability at Briansville
III	32 OZ Bottle capability at Vicksburg
IV	32 OZ Bottle capability at Madison

EVALUATING DESCRIPTION

A	Almost Perfect	O	Ordinary Results
E	Especially Good	U	Unimportant Results
I	Important Results	X	Not Acceptable

Values: A = 4, E = 3, I = 2, O = 1, U = 0, X = -1

COST DIFFERENCE FROM BASELINE

SAVINGS	INCREASED COSTS
-26% to -50%	0% to 5%
-16% to -25%	6% to 15%
-6% to -15%	16% to 25%
0% to -5%	26% to 50%

SUMMARY

1. Alt. I Jonesville is best in terms of cost.
2. From the intangible perspective Alt. II Briansville is best.
3. All factors considered, the new 32-oz. line for big bottles should be at Briansville.

... STEP 5 (continued)

COST COMPARISONS

Naturally, the planner must know when setting up the baseline model in Step 2 which costs will be of interest during evaluation in this Step 5. When comparing the operating costs of different logistical networks, the following are typically helpful:

- Raw material – including procurement transaction (ordering) costs and delivery to dock (freight).
- Inbound logistics – including receiving, inspection and put away into purchased components or raw material storage.
- Manufacturing – direct, variable overhead and relevant indirect expenses such as changeovers.
- Outbound logistics (Replenishment) – transportation costs from plant to warehouse, including the impact of any off-setting backhauls or reverse logistics.
- Storage costs – warehouse occupancy costs, including rents and infrastructure such as information systems.
- Warehousing activity costs – labor and operational expense
- Distribution costs – shipping, transportation, or delivery costs from finished good warehouse to customers, including the impact of any off-setting backhauls or reverse logistics.
- Inventory – carrying cost (including taxes) and any obsolescence.
- Depreciation – on manufacturing, warehousing, and transportation equipment and facilities.
- Taxes – that may change based on the jurisdictions where activities occur.

COST JUSTIFICATION

Typically, logistical network modeling tells us the annual operating cost for each network plan. If making some change to the network is a foregone conclusion, and the cost of making the change is about the same for all plans, then comparing operating costs is sufficient to decide which plan is best from an economic perspective. This is the case in our MTT example, where one production line will be upgraded and that cost will be about the same at any of the six plant locations. Management is simply asking which plant is best.

But if the question is whether to upgrade an existing line or build a new one, or whether to upgrade at all, then we must also decide if the projected savings or additional profit provide a sufficient return on the company's investment. This cost justification is in addition to our cost comparison among alternative plans. Relevant investment and one-time expense costs typically include:

- Facilities construction or modification
- Equipment – machinery, material handling, transportation, containers…
- Inventory – one time investments or draw downs (of not addressed through carrying costs above)
- Relocation or moving expense; installation expense

In our How-To instructions on Page 5-1, we note that annualizing investments and one-time expenses and adding these to operating costs gives a simple comparison of plans, and may reveal the most economical plan. But this still begs the question of whether any investment should be made. To answer this question, network planners should work with their associates in accounting and finance and apply the appropriate methods and procedures of cost justification. These are beyond the scope of this booklet.

EVALUATION OF INTANGIBLE FACTORS

Often, the proposed network plans are very similar in terms of operating costs and several of the alternatives may be cost justified. Yet, these plans may differ significantly in terms of their operational advantages or their ease of implementation. SNP calls these hard-to-quantify differences "intangibles" and evaluates them using the weighted-factor procedure explained and illustrated on Pages 5-1 and 2.

Cost-justified plans that are close in operating cost are usually chosen for their performance on intangible factors. Typical factors include:

- Ease of implementation (make the planned changes without disruption)
- Ease of management (ability to execute, operate or manage the network)
- Fit with existing organizational structure
- Labor issues such as exposure to strikes or high labor turnover
- Flexibility in the face of unforeseen changes in demand or product mix
- Equipment or capacity utilization and impact performance during peaks
- Space or real estate utilization
- Avoidance of risks of various kinds
- Image to the public or customer base
- Government regulations that may restrict or impede operations

WHAT MAKES A WINNER

Decision-making based on network modeling always involves assumptions and inferences about the future and the future is always uncertain. So, where costs estimates are within 10%, the economic results should be considered a toss up, and the selection should then be based exclusively on the intangibles comparison.

Given comparable economics, a 20% difference based on intangibles always makes a winner. If less than 5%, reconsider both the costs and the intangibles, perhaps involving different people.

In our MTT example, Jonesville's costs (Alt. I) are the lowest and would have saved about $700,000 per year over the current or baseline network. The remaining locations would have saved between $391,000 and $581,000. These savings easily justify the upgrade of a production line. But their economic differences are insignificant – only about 1% -- on total costs of $63 million per year. When intangible factors are considered, the lowest cost alternative (Jonesville) is ruled out because it rates "X" on risk of line shutdown. Of the remaining alternatives, Briansville scores highest at 84. It offers more capacity relief and easier implementation and its savings are somewhat greater than Vicksburg or Madison. For these reasons, MTT selected Briansville for the 32-ounce bottling line upgrade

HIDDEN FACTORS

Many proposed network changes involve sensitive personal issues, including facility shut-downs and layoffs, family relocations, reassignments and loss of responsibilities... Proposed changes may also have impact on unstated or still-secret business plans known only to a few decision-makers. These cannot be revealed but may color the discussion and influence the ultimate selection. Or, proposed plans may be at odds with positions already announced by key decision-makers, or their past experience with similar proposals.

The planner should take note when such factors may be present and tread carefully through the evaluation step. By doing the best job possible on cost comparison, cost justification and evaluation of intangibles, the planner can hopefully draw the decision-makers into an objective and communicative frame of mind and reduce or at least manage the influence of hidden factors.

OUTPUT OF STEP 5

> The output of Step 5 is **THE SELECTED NETWORK PLAN**.

Detail

And Do

STEP 6: DETAIL AND DO

In this final step we make detailed plans, take action and audit the success of our plan.

WHAT YOU DO

- Make a list of things that must be done to implement the selected plan.
- List these as tasks on the project plan and schedule sheet.
- Schedule the tasks and assign who is responsible for each.
- Periodically record the status of the scheduled tasks and take appropriate actions.
- Conduct a post-implementation audit and record the lessons learned.

Key Document = *Detail and Do Worksheet*

Output = *Implemented plan and audited results*

WHY YOU DO IT

Only by implementing our plans do we achieve results. And by auditing the results actually achieved, we learn how to improve our future modeling and planning.

HOW YOU DO IT

1. Fill in the headings at the top of the form. "Covering" defines the extent of the work to be done, "Status as of" and "Reported by" are used for control, when making status reports on the work being done.

2. Determine a calendar time scale for each vertical line on the right side of the form. Enter periodic dates at the top.

3. List each task to be accomplished in the "Work to do;Action to take" section. Enter one task per line in approximate chronological sequence.

4. Show who is responsible for doing each task.

5. Show the time of "earliest start date" and of "latest completion date" by brackets openings to the right and closing to the left. Show "duration" with a line connecting the two brackets.

6. As an optional addition, show the person-days (as distinguished from calendar days) by a number (of days) above the right end of each duration bracket.

7. Use the project plan and schedule to stay on schedule. Weekly or at regular intervals, place a V (representing an arrowhead) at the date of the review on the top calendar scale. Then fill in for each task, with a heavy marking line, the portion of that task completed, starting at the left bracket. Thus each task with its percent complete line falling beyond the V-marked point is ahead of the schedule and each heavy line not extending to that date is behind schedule.

   As noted above, when reporting status to others, after marking the portions of tasks complete, enter the "Status as of" date. Enter the initials of "Reported by" and distribute to the appropriate parties.

8. A suitable time – typically 90 days to one year – after implementation is complete, review the network's actual costs and compare them to those projected by the network model. Explain the variances note the lessons learned for future projects.

DETAIL AND DO

Covering	Manufacturing, Logistics, Planning	Company	MTT	Project	SNP-106	By	CN
	KW, BF, RH, CN, KL, DS,TB	Status as of	10/13	Project Description	Briansville Line Upgrade	With	DS, RB, TB
Distribution		Reported by	CN	Date	9/23	Sheet 1 of 1	Further

Task/Proj. No.	Work to do: Action to take	Resp. of	Days after start: 1 2 3 4 5 6 7 8 9 10 11 12 13 14 15 16 17 18 19 20 21 22 23 24 25	Further Schedule
1.	Communicate the selection plan and schedule upgrade date	TS		
2.	Perform the vendor selection for change parts	MS		
3.	Perform the line upgrade	MS		
4.	Instruct the planning department	CN		
5.	Allocate the demand to the manufacturing location	MM		
6.	Inform the SKU committee of the changes	SB		
7.	Set up the sourcing in the system to reflect change	CW		
8.	Make changes to the ERP system to reflect the new cost	CW		
9.	Communicate the changes to logistics	DS		
10.	Change the raw material sourcing	DC		
11.	Commit the changes and begin production	TS		
12.				
13.				
14.				
15.				

POST IMPLEMENTATION AUDIT

COST SUMMARY

No.		UOM	Projected Savings	Actual Savings	Variance	Notes
1.	Manufacturing costs	$	$0	$18,000	$18,000	1.
2.	Transportation costs	$	$535,000	$369,000	($166,000)	2.
3.	Warehousing labor costs	$	$0	($6,500)	($6,500)	3.
4.	Storage costs	$	$0	$13,000	$13,000	4.
5.	Reduced overtime	$	$28,000	see Mfg. Cost above	($15,100)	5.
6.	Raw Material Procurement costs	$	$0	$52,000	$52,000	6.
7.	Cross docking costs	$	$10,800	$10,000	($800)	7.
8.						
9.						
10.						

NOTES AND EXPLANATION AND LESSONS LEARNED

No.	
1.	Reduced overtime at Sommersville is credited as a reduction in manufacturing cost. Projection was for $15,000. Actual reduction was $18,000.
2.	The full realization was not obtained because of steep increase in the fuel costs. Consider fuel price projections in future models.
3.	Warehousing direct labor costs increased because of increased labor activity.
4.	Storage costs on a per unit reduced because of better absorption of the fixed costs. But trivial in terms of total cost. Correct to leave out of the model.
5.	Overtime at Sommersville was eliminated as planned but is treated as a reduction in manufacturing cost on line 1.
6.	Raw material costs reduced because MTT negotiated a contract for producing all of its big bottles in Briansville and raw material plant was in close vicinity to Briansville.
7.	Total savings not significant and it was correct to leave this out of the cost evaluation.
8.	Cross docking costs were 8% higher than expected but still insignificant in terms of total cost.

☐ Date work scheduled to start ☐ Date work scheduled to finish ☐ Total time scheduled for work
— Amount of work done ▼ Reporting indicator (Each vertical period represents one unit of time)

© Copyright 2007. CHANDRA NATARAJAN AND RICHARD MUTHER & ASSOCIATES - 7/68

May be reproduced for in-company use provided original source is not deleted.

6-2

... STEP 6 (continued)

WORKING OUT THE DETAILS

The network plan selected in Step 5 will usually still have some details to be worked out. And a final approval may still be needed before implementation can begin. The planner should first verify acceptance of the selected plan and get formal sign-offs. This may involve verifying total cost and availability of funds.

Working through the remaining details typically involves finalizing for facility and item (or stock keeping unit):

- Which customers will be served and from which locations?
- What products will be supplied and from where?
- Which products will be made internally and which sourced from outside?
- Which products will be made or distributed at which locations, and in what quantities?
- How much capacity will be provided at each producing or distributing location?
- Which suppliers will be used for which locations?
- How much inventory will be held at which locations?
- What will be the hours and days of operation?
- What modes of transportation will be used?

The implementation planning cannot be completed without these details.

IMPLEMENTATION ACTIVITIES

Up to now we have been making plans – analyzing, selecting and then detailing the best network plan. Now we must identify and clarify the actions needed to implement the plan. We do this by making a list of things to do, assigning who is responsible, setting a schedule and making sure that things get done.

Often, the network planner's role ends with the recommendation. But it should not. Participating in the implementation planning and the implementation itself improves the planners' understanding of the realities and practicalities of supply chain and logistical network management. Thus, we become better modelers on future projects.

Typically, implementation requires the following kinds of actions:

   Administrative actions – easy changes, quick to make
   - Changes to planning systems
   - Addition of SKU's into system
   - Adding cost centers to capture new products
   - Forecasting for new products and locations
   - Addition or deletion of steps in the systems

   Operational actions – more difficult changes and time-consuming
   - Creating new transport lanes
   - Producing at new locations
   - Warehousing at new locations
   - Buying from new locations

   Infrastructure actions – generally long lead-times and large-scale effort
   - Identifying a new location
   - Building or leasing a new facility
   - Buying or leasing new material handling equipment
   - Hiring labor
   - Moving to a new facility

Administrative and operational changes can be made more easily than those to infrastructure and should be made first. As these and the infrastructure changes are made, the planner's role shifts to the important one of project manager, communicating the status of changes to the appropriate people and functional groups.

If there is resistance to making the necessary changes, the planner should share and re-state the Step 4 and Step 5 results, explaining why the project is beneficial to the company as a whole while recognizing that the impact of implementation may affect some functions more than others, and the impact may be negative for some.

POST IMPLEMENTATION AUDIT

The post implementation audit compares actual project savings to what was expected for the plan. This helps to understand any variances from the projected savings and the reasons therefore. This understanding can help improve the accuracy of future models. It will also confirm the correctness of assumptions and the validity of parameters and constraints. The audit should also review the outcomes with respect to intangible considerations in Step 5.

Our example from MTT shows the benefits of running such an audit. Fuel-price increases eliminated half the projected savings. Given this impact the planners would be wise to include fuel price projections in future models of this type. Other variations were insignificant.

OUTPUT OF STEP 6

> The output of Step 6 is an **IMPLEMENTED PLAN AND AUDITED RESULTS**.

CASE EXAMPLES

On the sheets immediately following are three case examples. Each shows the six steps and working forms of Simplified SNP as applied to solve various problems.

CASE EXAMPLE 1
MTT

BIG BOTTLE ANALYSIS

Example One – Big Bottle Analysis shows the six steps for Mountain Trail Tonic. This is the same carry-through example, shown here all together, that we used on page 2 of each of the six chapters. The planners are using an existing network model to decide the best place to add capacity for 32-ounce bottles among six existing plants. The network *model* compares costs for manufacturing, transportation, cross-docking and purchasing. The planning *project* considers both modeled costs and intangible factors such as ease of implementation, organizational impact, risk of shut down, and physical condition of the lines involved.

CASE STUDY 1 MONTAGE

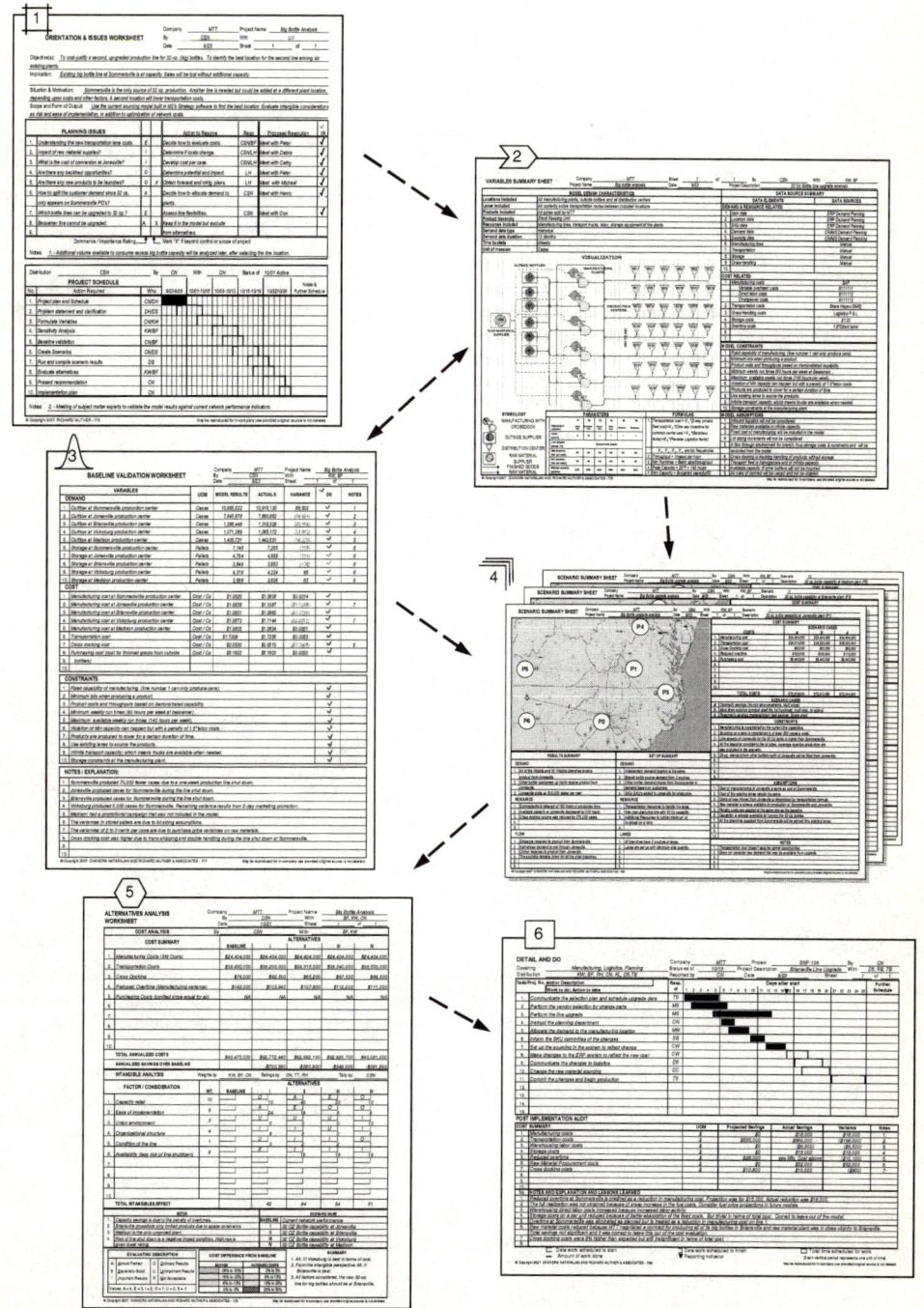

CASE EXAMPLE 2
JUST SPRINGS

CONTINGENCY MODEL

Example Two – Tampa Bay contingency plan shows the six steps applied to inventory planning at Just Springs Water. Each year during hurricane season, this company loses production at its Tampa Bay, Florida plant. The planners are using an existing network model to evaluate various ways to maintain supply and customer service. These include serving the area from other plants, running overtime and pre-building inventory, the use of a contract bottler and the use of a third party warehouse to hold pre-built stock. In addition to modeled network costs, important intangible considerations include: impact on customer service; fit with the strategic desire to use of third-party logistics; and the ability of the chosen plan to be used on an annual basis (repeatability).

ORIENTATION & ISSUES WORKSHEET

Company	Just Springs Water
By	RM, SN
Date	1/23
Project Name	Tampa Bay contingency plan
With	Sales, Operations, Logistics
Sheet	1 of 1

Objective(s): Maintain supply to Florida Region DCs and control costs during Tampa hurricane season shutdowns. Maintain customer service.

Implication: May need to serve Florida region from other plants in Southeast, or prebuild inventory in Tampa and hold, or use contract bottler. Potential impact on costs and service. Approaches differ in risk and ability to execute. Using other plants or prebuild will require overtime production.

Situation & Motivation: Typical season results in 2 weeks' lost production, 50% stockouts, and 200% cost overrun on make-up sales. Could serve Florida from Greenville, Jackson, Knoxville, or contract bottler. Need to find lowest cost solution subject to customer service and other factors.

Scope and Form of Output: Use current Southeast Region sourcing model. Prepare written report and presentation to Sales, Planning and Logistics managers.

	PLANNING ISSUES			Action to Resolve	Resp	Proposed Resolution	✓ ok
1.	How much capacity can be tapped at other plants?	E	(1)	Check capacities at Southeast plants.	SN, RM	Meet with planners.	✓
2.	How much can be stored at each plant and DC?	E	(1)	Check storage capacities.	TM	Meet with Henrico.	✓
3.	What are the transportation costs from each plant?	A		Determine the costs.	TK	Get estimates from Logistics.	✓
4.	What capacity is available from outside suppliers?	I		Identify the preferred contract bottler and determine their capacity.	ST	Meet with purchasing.	✓
5.	What is the cost of buying from outside suppliers?	I		Get quotes from contract bottlers.	SN	Purchasing to get quote.	✓
6.	Who will transport product from remote plants?	I		Identify the transportation carrier.	MS	Meet with Tom.	✓
7.	What are the overtime costs for manufacturing?	O		Determine overtime costs.	MS	Meet with Rui.	✓
8.	Do we have yard capacity to hold prebuild product?	O	(1)	Determine yard space at each site.	MS	Meet with Tiffany.	✓
9.	Can rail carrier MDX hold prebuild in rail cars?	E		Check with MDX Rail.	MS	Work through Tom.	✓

Dominance / Importance Rating ↑   ↑ Mark "X" if beyond control or scope of project

Notes: (1) Plan must work within available capacities at plants, DCs and yards.

Distribution	SN, RM, TM, TK, ST, MS	By	SN	With	RM	Status of	01/29

PROJECT SCHEDULE

No.	Action Required	Who	01/21-01/25	01/28-02/01	02/04-02/08	02/11-02/15	02/18-02/22	Notes & Further Schedule
1.	Project schedule	SN	■					
2.	Problem statement and issues clarification	RM	■	■				
3.	Formulate variables	SN		■				
4.	Analyze sensitivities	SN		■				
5.	Baseline validation	SN			■			
6.	Create scenarios	RM, MS, SN			■			(2)
7.	Run and compile scenarios	SN				■		
8.	Evaluate alternatives	RM, MS, SN				■		
9.	Present recommended plan, approve & implement	RM, MS, SN					■	Implement by 06/01
10.								

Notes: 2. Need quote from contract bottler by 2/10

© Copyright 2007. RICHARD MUTHER - 773

VARIABLES SUMMARY SHEET

Company	Just Springs Water	Sheet	1 of 1	By	RM, SN	With Sales, Operations, Logistics
Project Name	Tampa Bay contingency plan	Date	2/1	Project Description	Florida DC sourcing plan during hurricane season shut downs	

MODEL DESIGN CHARACTERISTICS

Locations included	Raw material suppliers' plants, outside suppliers' plants, Southeast Region plants and all SE Region DCs
Lanes included	All existing lanes between raw material suppliers, production facilities and DCs
Products included	All products made and distributed by Just Springs Water
Product hierarchy	Brand level for every location (stock keeping unit)
Resources included	Production lines, transportation, storage and handling, warehousing
Demand data type	Historical (Use last hurricane season)
Demand data duration	8 Months
Time buckets	Weekly
Unit of measure	Cases

NETWORK DIAGRAM

(Network diagram showing flows from raw material suppliers through plants in Greenville SC, Jackson MS, Knoxville TN, Tampa Bay FL to distribution centers DC1–DC14.)

PARAMETERS

Plant Name / Characteristics	Knoxville, TN	Greenville, SC	Jackson, MS	Tampa Bay, FL
Number of Lines	2	2	2	2
Speed of Line	Demonstrated Speed	Demonstrated Speed	Demonstrated Speed	Demonstrated Speed
Maximum Capacity	112 Hrs / Wk	112 Hrs / Wk	112 Hrs / Wk	112 Hrs / Wk
Minimum Capacity	80 Hrs / Wk	80 Hrs / Wk	80 Hrs / Wk	80 Hrs / Wk
UOM	Hrs	Hrs	Hrs	Hrs
Pallets Stored	5000	6000	4500	6000

FORMULAS

1. Overtime = 1.5 * direct labor costs
2. Min run time = batch size / throughput
3. Max capacity = 112 Hrs/Wk
4. Transportation cost = $X1$*(2-way private fleet cost)+$X2$*(One way incentive for common carrier use) +$X3$*(Backhaul factor)+$X4$*(Reverse Logistics factor)
5. $X1, X2, X3, X4$ are trip frequencies
6. Throughput = 1 / Cases per hour

SYMBOLOGY

- MANUFACTURING WITH CROSSDOCK
- OUTSIDE SUPPLIER
- DISTRIBUTION CENTER
- RAW MATERIAL SUPPLIER
- FINISHED GOODS
- RAW MATERIAL

DATA SOURCE SUMMARY

DATA ELEMENTS | DATA SOURCES

DEMAND & RESOURCE RELATED

	DATA ELEMENTS	DATA SOURCES
1.	Item data	Product master file
2.	Location data	Oracle ERP system
3.	Stocking Keeping Unit	Oracle ERP system
4.	Demand data	Oracle S & D system
5.	Move(lane) related data	Sourcing plan (move summary)
6.	Production resources	PP systems
7.	Transportation resources	TMS systems
8.	Storage resources	Manual
9.	Cross handling resources	Manual
10.		

COST RELATED

1.	Production costs	Oracle finance M & L cost center
2.	Labor costs	Oracle finance M & L cost center
3.	Overtime Costs	Oracle finance M & L cost center
4.	Storage costs	Oracle finance WH cost center
5.	Warehousing Labor costs	Oracle finance WH cost center
6.	Warehouse overtime costs	Oracle finance WH cost center
7.	Handling costs	Oracle finance WH cost center
8.	Raw material costs	Oracle purchase
9.	Transportation costs	TMS
10.		

MODEL CONSTRAINTS

1. Raw material supplier locations are considered as sources with fixed capacity.
2. Available manufacturing capacity is fixed.
3. Transportation capacity is infinite and all lanes are included in the baseline.
4. Production is based on minimum run requirements.
5. Maximum available capacity is 112 hrs/wk & 96 hrs/wk in a union environment.
6. Demand at distribution centers must be met.
7. Storage capacities of DCs are fixed at current capacities.
8. Every package has different minimum run times at different plants.
9. Dummy lanes are used between plants and Florida DCs during hurricanes.
10.

MODEL ASSUMPTIONS

1. Every raw material supplier sends a fixed percentage of required supply.
2. Continuous production is assumed and changeovers are built into speeds.
3. Use demonstrated line speeds for each package type.
4. Backhaul and reverse logistics costs are built into the transportation costs (formula 4).
5. Additional available capacity of the contract manufacturers will not be considered.
6. Model assumes a standard pallet cube for all products.
7. Labor resources are always available as overtime and temps but at a premium.
8. Uptimes of every line will be controlled using the line speeds.
9. Every package change takes 40 minutes to set up.
10.

© Copyright 2007. CHANDRA NATARAJAN AND RICHARD MUTHER & ASSOCIATES - 775

May be reproduced for in-company use provided original source is not deleted.

BASELINE VALIDATION WORKSHEET

Company: Just Springs Water
By: SN, RM
Date: 2/5
Project Name: Tampa Bay contingency plan
With: TW, DH
Sheet: 1 of 1

VARIABLES	UOM	MODEL RESULTS	ACTUALS	VARIANCE	OK	NOTES
DEMAND						
1. Greenville plant production	Cases	8,321,092	8,436,552	(115,460)	✓	1.
2. Tampa plant production	Cases	7,965,780	7,534,642	431,138	✓	2.
3. Jackson plant production	Cases	8,395,752	8,565,657	(169,905)	✓	3.
4. Knoxville plant production	Cases	6,425,768	6,425,768	0	✓	
5. Naples inventory	Pallets	2,795	2,978	(183)	✓	4.
6. Tampa inventory	Pallets	8,997	8,676	321	✓	5.
7. Ocala inventory	Pallets	2,568	2,741	(173)	✓	5.
8. St. Petersburg inventory	Pallets	2,578	2,735	(157)	✓	5.
9. Sarasota inventory	Pallets	2,006	2,167	(161)	✓	5.
10. Purchased cases from outside bottler	Cases	1,637,056	1,782,829	(145,773)	✓	6.
COST						
1. Greenville production	Cost / Cs	$1.18	$1.19	($0.0070)	✓	
2. Tampa production	Cost / Cs	$1.21	$1.18	$0.0266	✓	7.
3. Jackson production	Cost / Cs	$1.19	$1.20	($0.0087)	✓	
4. Knoxville production	Cost / Cs	$1.21	$1.21	$0.0000	✓	
5. Total average warehousing storage	Cost / Cs	$0.17	$0.17	$0.0005	✓	
6. Total average warehouse labor	Cost / Cs	$0.22	$0.22	($0.0048)	✓	8.
7. Total average transportation	Cost / Cs	$0.57	$0.56	$0.0125	✓	9.
8. Price of finished goods from outside supplier	Cost / Cs	$6.63	$6.63	$0.0000	✓	
9.						
10.						

CONSTRAINTS

		OK	
1.	Raw material supplier locations are considered as sources with fixed capacity.	✓	
2.	Available manufacturing capacity is fixed.	✓	
3.	Transportation capacity is infinite and all lanes are included in the baseline.	✓	
4.	Production is based on minimum run requirements.	✓	
5.	Maximum available capacity is 112 hrs/wk & 96 hrs/wk in a union environment.	✓	
6.	Demand at distribution centers must be met.	✓	
7.	Storage capacity of DCs are fixed at current capacities.	✓	
8.	Every package has different minimum run times at different plants.	✓	
9.	Dummy lanes are used between plants and Florida DCs during hurricanes.	✓	
10.			

NOTES / EXPLANATION:

1. Greenville produced 116,000 cases for the Florida DCs during the hurricane season.
2. Tampa plant was shut down for 8 days due to hurricanes which reduced its production.
3. Jackson produced 169,900 cases for Florida DCs during the Tampa hurricane shut downs.
4. Naples, St. Petersburg, Ocala and Sarasota stored pallets during Tampa shut downs normally stored at Tampa hub DC.
5. Tampa stored 300 fewer pallets on the average. This was due to the 8 days production shut down Tampa had during hurricane
6. Outside supplier produced 150,000 cases for Florida DCs during the plant hurricane shut down.
7. Production cost variance is due to line shut down at Tampa for 8 days during the hurricane season.
8. Incremental costs experience by warehouse due to hurricanes. Costs include temps, OT, OT benefits.
9. Incremental costs experienced by transport due to hurricanes. Costs include new temporary lanes used between plants and Florida DCs during Tampa shut downs.

© Copyright 2007. CHANDRA NATARAJAN AND RICHARD MUTHER & ASSOCIATES - 777   May be reproduced for in-company use provided original source is not deleted.

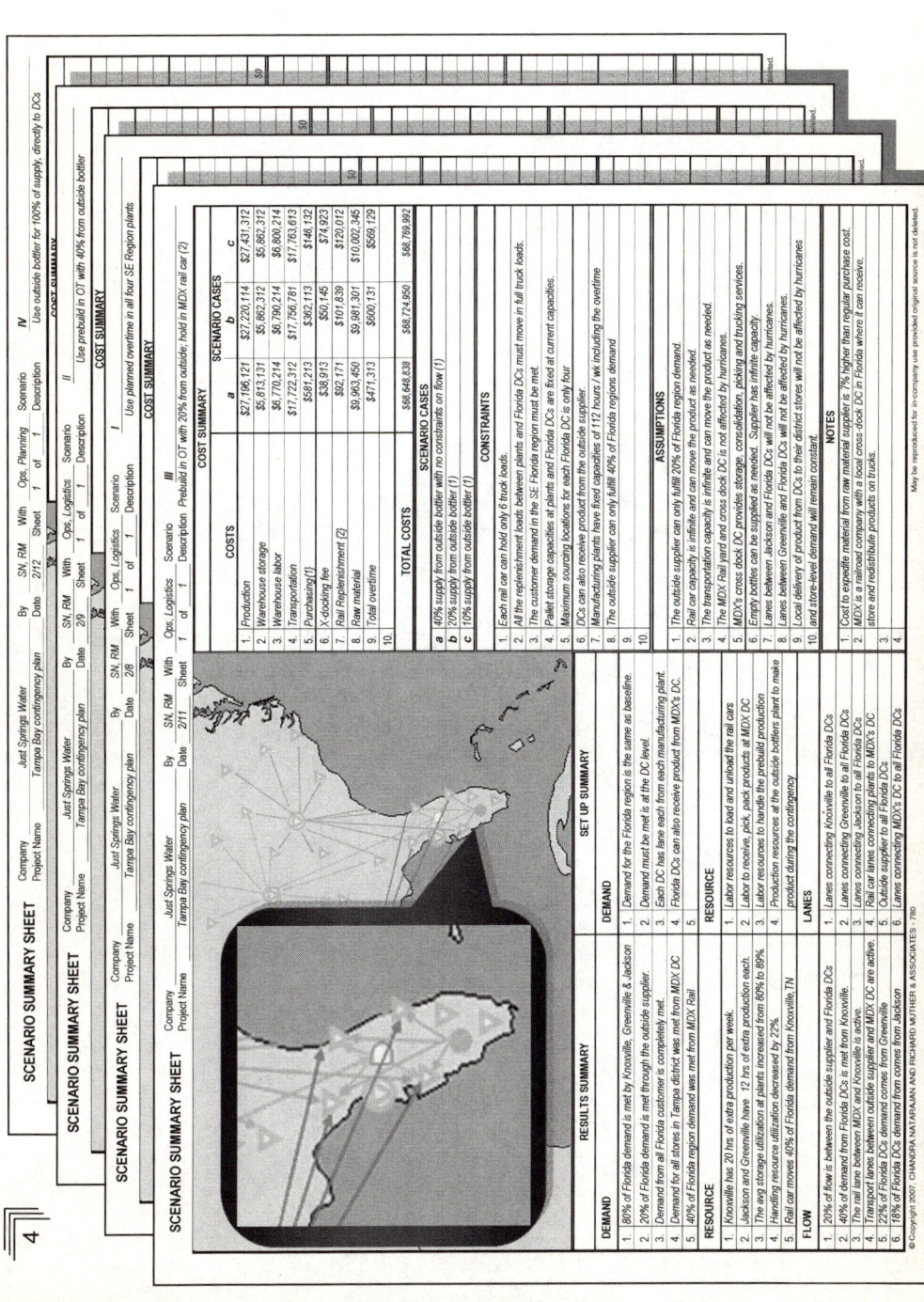

ALTERNATIVES ANALYSIS WORKSHEET — 5

Company	Just Springs Water
By	RM, SN, DH
Date	2/15
Project Name	Tampa Bay contingency plan
With	Sales, Operations, Logistics
Sheet	1 of 1

COST ANALYSIS

By: TW, RM, SN, DH — With: Finance

	COST SUMMARY	BASELINE	I	II	III	IV
1.	Production costs	$27,321,140	$27,121,140	$27,112,345	$27,220,114	$26,944,234
2.	Warehouse storage costs	$5,986,321	$6,013,104	$5,972,346	$5,862,312	$5,913,124
3.	Warehouse labor costs	$6,534,621	$6,581,312	$6,634,523	$6,490,214	$6,590,213
4.	Transportation costs	$17,867,432	$17,811,919	$17,756,791	$17,756,781	$17,734,153
5.	Purchasing costs {1}	$0	$0	$565,342	$362,113	$1,052,024
6.	Cross-docking fees on cases via MDX Rail DC	$0	$0	$0	$50,145	$0
7.	MDX Rail transport and holding costs {2}	$0	$0	$0	$101,839	$0
8.	Raw material costs	$10,245,859	$10,101,404	$9,909,654	$9,981,301	$9,844,513
9.	Overtime costs at plants and warehouses {3}	$1,221,091	$1,037,301	$704,560	$600,131	$388,456
10.						
	TOTAL ANNUALIZED COSTS	$69,176,464	$68,666,180	$68,655,561	$68,424,950	$68,466,717
	ANNUALIZED SAVINGS OVER BASELINE		-$510,284	-$520,903	-$751,514	-$709,747

INTANGIBLE ANALYSIS

Weights by: RM, JD, BF — Ratings by: TW, RM, SN, DH, MS — Tally by: SN

	FACTOR / CONSIDERATION	WT.	BASELINE	I	II	III	IV
1.	Ability to execute	6		E 18	I 12	I 12	O 6
2.	Impact on customer service {4}	10		I 20	I 20	A 40	I 20
3.	Impact on labor availability	2		U 0	E 6	I 4	A 8
4.	Ability to use available system capacity	7		I 14	U 0	E 21	U 0
5.	Potential impact on salaried staffing	3		U 0	O 3	I 6	E 9
6.	Ease of plan communication	5		I 10	U 0	O 0	U 0
7.	Ability to repeat the plan (use every year)	8		U 0	O 8	I 16	O 8
8.	Fit to strategic vision {5}	9		U 0	I 18	E 27	X
9.	Dependability of plan	4		E 12	I 8	I 8	O 4
10.							
	TOTAL INTANGIBLES EFFECT			74	75	134	55

NOTES

1. Cost incurred to fill demand from an outside bottler.
2. Cost of using MDX Rail for prebuild is in addition to transportation costs.
3. Total overtime costs incured by the Logistics department.
4. Continuity of supply. Overtime and outside bottler options may be less dependable.
5. Just Springs strategic planning envisions using rail and 3rd party logisitics for Florida.

SCENARIO NAME

BASELINE	Current operation with no contingency plan
I	Use planned overtime in all four Southeast Region plants
II	Use prebuild in OT with 40% from outside bottler
III	Prebuild in OT with 20% from outside; hold in MDX rail car.
IV	Use outside bottler for 100% of supply, directly to DCs

SUMMARY

1. Scenario IV is the best plan from a cost perspective.
2. Scenario III is best based on intangibles.
3. Scenario I uses a planned overtime and is the highest cost plan
4. Based on both cost and intangibles, Scenario III is the best contingency plan for Tampa Bay.

EVALUATING DESCRIPTION

A	Almost Perfect	O	Ordinary Results
E	Especially Good	U	Unimportant Results
I	Important Results	X	Not Acceptable

Values: A = 4, E = 3, I = 2, O = 1, U = 0, X = -1

COST DIFFERENCE FROM BASELINE

SAVINGS	INCREASED COSTS
-26% to -50%	0% to 5%
-16% to -25%	6% to 15%
-6% to -15%	16% to 25%
0% to -5%	26% to 50%

© Copyright 2007. CHANDRA NATARAJAN AND RICHARD MUTHER & ASSOCIATES - 178
May be reproduced for in-company use provided original source is not deleted.

DETAIL AND DO WORKSHEET

Covering	Planning, Operations, Logistics, Purchasing, Human Resources
Distribution	RM, SN, TW, DH, TF, RHH, JJB
Company	Just Springs Water
Status as of	9/22
Reported by	RM
Project	Tampa Bay contingency plan
Project Description	Contingency plan for Tampa region
By	RM, SN, PW
With	BF
Sheet	1 of 1

Task/Proj. No. and/or Description	Resp. of	8/22-8/27	8/29-9/03	9/06-9/10	9/13-9/17	9/20-9/24	Further Schedule
		Date 8/17				▼	
1. Create locations in the planning system: MDX DC	RM	■					
2. Create SKU's in planning system for contingency plan	SN	■					
3. Create sourcing plan for contingency and flag it to use it for hurricanes	RM	■					
4. Create lanes in the TMS system to indicate contingency plan	TW		■				
5. Create raw material supply schedule for prebuild	TF		■				
6. Schedule overtime for planned internal prebuild	RHH		■				
7. Plan the staffing for the internal prebuild	TT			■			
8. Negotiate outside bottler agreement	SN			■			
9. Negotiate MDX Rail contract for cross docking	SN			■			
10. Schedule overtime for planned internal prebuild	RM				■		
11. Contract with the common carriers to move the product during hurricane	TF				■		
12. Make a visit to MDX DC	JJB					■	
13. Contingency plan manual	RHH					■	
14.							
15.							

POST IMPLEMENTATION AUDIT (1)

COST SUMMARY	UOM	Projected Savings (1)	Actual Savings	Variance	Notes
1. Production costs	$	$63,141	$51,300	($11,841)	2
2. Warehouse storage costs	$	$77,506	$60,000	($17,506)	3
3. Warehouse labor costs	$	$28,496	$20,000	($8,496)	3
4. Transportation costs	$	$69,157	$50,000	($19,157)	4
5. Purchasing costs	$	($226,321)	($150,000)	$76,321	5
6. MDX Rail cross docking fee	$	($31,341)	($40,000)	($8,659)	
7. Rail replenishment costs	$	($63,649)	($60,000)	$3,649	
8. Raw material costs	$	$165,349	$150,000	($15,349)	6
9. Total overtime costs (warehouse, production)	$	$388,100	$360,000	($28,100)	7
10.					

NOTES AND EXPLANATION AND LESSONS LEARNED

No.	
1.	Hurricanes caused the Tampa line to be shut down for only 5 days. Projected savings for an expected 8 days of shut down has been prorated to match 5 days of savings.
2.	Production costs were higher expected because Just Springs was able to build more than the planned 80% of volume.
3.	The warehouse storage and labor costs increased because Just Springs plants -> Jackson, Knoxville, Greenville produced more volume.
4.	Transportation savings was lower than expected because of a 20% increase in fuel costs.
5.	Purchasing costs were lower than projected since the outside bottler delivered less than the planned 20% of volume.
6.	Raw material costs were higher than expected due to increases in inbound trucking costs and prebuild was a higher percentage of plan.
7.	Total overtime costs were higher because Just Springs plants built more cases than the planned 80% of volume.
8.	
9.	

☐ Date work scheduled to start
■ Amount of work done
☐ Date work scheduled to finish
▼ Reporting indicator
☐ Total time scheduled for work
(Each vertical period represents one unit of time)

© Copyright 2007. CHANDRA NATARAJAN AND RICHARD MUTHER & ASSOCIATES - 768

May be reproduced for in-company use provided original source is not deleted.

CASE STUDY 2 MONTAGE

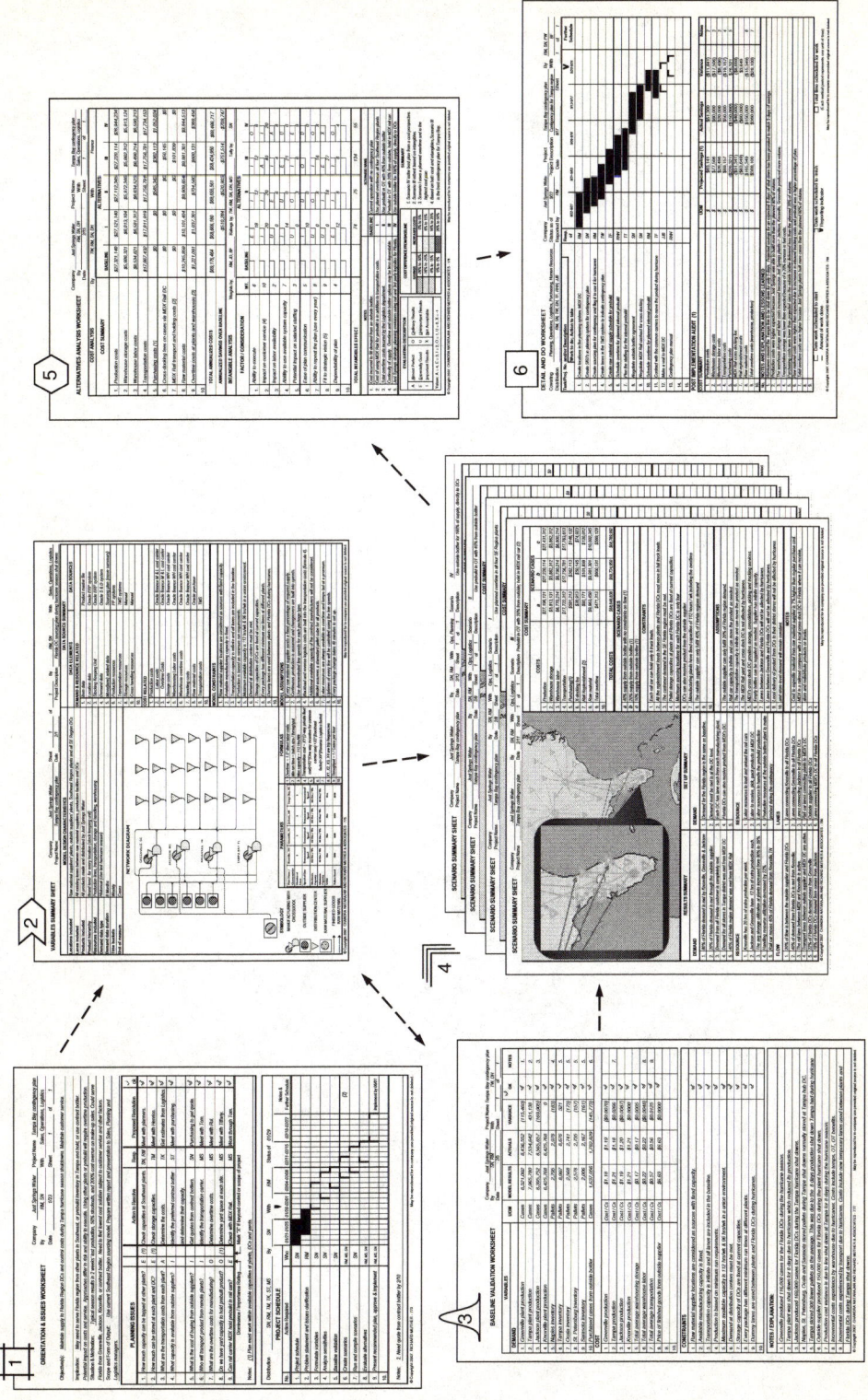

CASE STUDY 3
KMG EXAMPLE

DC OPERATION'S STRATEGY

Example Three – LaGrange Operations Strategy shows the six steps applied to strategic planning. The KMG company has too many warehouses. The planners are deciding if LaGrange should remain as a distribution center, be reduced to a cross-dock or drop lot, or close altogether with its customers served by other existing warehouses. Outsourcing is also being considered. In addition to modeled network costs, the planners must also consider the company's ability to operate the chosen strategy, its acceptance by local sales management and impact on customer service.

ORIENTATION & ISSUES WORKSHEET

Company	KMG Inc.
By	LHH, SN, KP
Date	7/16
Project Name	LaGrange Operations Strategy
With	Sales, Ops, Supply Chain, Finance
Sheet	1 of 1

Objective(s): To identify the best operating strategy for the LaGrange distribution center. Continue as is or change to: Cross-dock?; Drop lot?; Close and serve from other DCs?; Outsource to 3PL? Select based on costs and intangibles.

Implication: If the lowest cost with acceptable service is from a 3rd party or another company location, then outsource or close. But the selected strategy must maintain good customer service.

Situation & Motivation: Recent acquisition resulted in too many warehouses in Georgia. This has resulted in too much inventory and overhead.

Scope and Form of Output: Scope of the model is the state of Georgia. Evaluate costs with current regional network model built in the CDA strategy software. Output is a written report and presentation to VPs of Sales, Operations, Supply Chain and Finance.

No.	PLANNING ISSUES	D/I	X	Action to Resolve	Resp	Proposed Resolution	ok
1.	What is the impact on inventory?	E		Decide how to calculate inventory	LHH	Use statistical model	✓
2.	What is the impact on overhead costs?	A		Calculate the salaried staff required	SN	Use KMG's ratio allocation	✓
3.	How to calculate warehouse labor required?	I		Determine the labor requirement	MM, SN	Agree on max. picking rates	✓
4.	How estimate the transportation costs?	E		Determine the costs per lane	MM, SN	Use cost / unit / lane formula	✓
5.	Are there any 3rd party warehouse available to store the product?	O		Consider only MTX's warehouses as available warehouses	SN	Meet with Cathy Cannon	✓
6.	How to estimate handling equipment required?	I		Calculate from pallets handled	SN	Confirm standards in model	✓
7.	How to model the impact on routes & frequency?	U		Use the current model	RL	Validate during modeling	✓
8.	Are there any community image concerns?		X	Discuss with top management	KL	Rate as intangible factors	1

Dominance / Importance Rating ↑   ↑ Mark "X" if beyond control or scope of project

Notes: 1. Intangibles do not rule out any strategy from quantitative analysis. Consider all possibilities.

Distribution	Sales, Operations, Supply Chain, Finance
By	SN, LHH
With	VP of SC
Status of	8/8

PROJECT SCHEDULE

No.	Action Required	Who	7/16-7/20	7/23-7/27	7/30-8/03	8/06-8/10	8/13-8/17	Notes & Further Schedule
1.	Project scheduling and timeline development	SN, LHH	■					
2.	Problem & issues --> clarification and resolution	SN	■					
3.	Define the problem variables	SN, LHH		■				
4.	Analyze the sensitivities	SN		■				
5.	Validate the current baseline performance	SN, LHH			■			2
6.	Define and create operating scenarios	SN, LHH			■			
7.	Run and compile scenario results	SN, LHH				■		
8.	Evaluate the alternatives	SN				■		
9.	Present recommendation to the operations team	SN, LHH					■	
10.	Detail and implement the selected plan	SN, LHH					■	Implement by Oct. 1

Notes: 2. Consensus meeting between the network planning team, Sales, Operations, Supply Chain and Finance to validate the baseline model

VARIABLES SUMMARY SHEET

Company	KMG	Sheet	1 of		By	LHH, SN, KP	With Sales, Ops, SC, Finance
Project Name	LaGrange Operations Strategy	Date	7/25		Project Description	Identify best strategy for operating LaGrange	

MODEL DESIGN CHARACTERISTICS

Locations included	Georgia-related locations: Outside suppliers; Mfg. plants; Georgia DCs; Georgia customers
Lanes included	All active lanes: Outside supplier-->plants and DCs; Plants-->DCs, DCs-->Customer stores
Products included	All candy and snack food products sold in Georgia
Product hierarchy	Modeling will be performed at the product level
Resources included	All Manufacturing lines, Warehouses, Handling & Storage equipment, Trucks, Labor
Demand data type	Combination of history and forecast data
Demand data duration	3 years
Time buckets	Weekly
Unit of measure	Units (carton of eaches)

DATA SOURCE SUMMARY

DEMAND & RESOURCE RELATED

DATA ELEMENTS	DATA SOURCES
1. Item Data	SNO systems
2. Location Data	SNO systems
3. Demand Data	ERP S&D, Routing files
4. Sourcing Data	SNO systems
5. Customer Clustering rule	Customer master file
6. Storage system resources	Manual
7. Replenishment and Distribution trucks	Manual
8. Real Estate resources	Manual
9. Handling resources	Manual
10. Labor resources	Manual

COST RELATED

1. Service cost	wt. avg costs derived from DP (1)
2. Distribution truck and fuel cost	Logistics P&L, 114
3. Distribution labor cost	Logistics P&L, 117
4. Replenishment cost	Transportation P&L, 215
5. Storage cost	Logistics P&L 310
6. Warehouse labor costs	Logistics P&L, 313
7. Inventory carrying costs	Logistics P&L, 314
8. Manufacturing costs	Production P&L, 10
9. Maintenance costs	Logistics P&L, 102
10. Fuel Surcharge	Mike McCain

MODEL CONSTRAINTS

1. Customer demand must be met.
2. Products will flow as per the current sourcing set up.
3. All lanes can only move full truck loads of product.
4. The storage capacity at each distribution center is fixed.
5. Each customer store must be served by only one DC
6. Total inventory at any point in any DC will be modeled as avg. inventory + safety stock.
7. Customers will be served at fixed frequencies to accurately depict service & dist. costs.
8. Distribution trucks will carry only the needed demand (based on route plan).
9. Warehouse labor will be available for 80 hrs each week.
10. Labor hours may not exceed fixed resources (e.g. LaGrange has 6 pickers).

MODEL ASSUMPTIONS

1. Trucks will carry only full truck load of product.
2. Full truck loads between plants and DCs are 28 pallets.
3. Transport trucks are always available.
4. Regressive cost model: $Y = Mx + C$ where C is fixed cost, M is cost per pallet, x is pallets.
5. Distribution trucks are always available.
6. DC replenishment is instantaneous. No lead time between need and receipt.
7. Raw material supply costs are assumed to be insignificant and not considered.
8. Line shut downs and maintenance will not considered and will be explained by variance from model results.
9. Outside suppliers will always send full truck loads of product.

VISUALIZATION

MANUFACTURING PLANTS: P1 Marietta, GA; P2 Jacksonville, FL; P3 Columbia, SC

DISTRIBUTION CENTERS: Albany, GA; Savannah, GA; Augusta, GA; LaGrange, GA

ANY TO ANY → CUSTOMER STORES

OUTSIDE SUPPLIERS: S1, S2, S3, S4

RAW MATERIAL SUPPLIER: RM

PARAMETERS

Characteristics	DC1 Albany	DC2 Augusta	P1 Marietta	DC3 LaGrange	P2 Jack'ville	P3 Columbia	DC4 Savannah
Storage Capacity (pallets)	25000	20000	9000	18000	8000		2000
Max Time available (hours)	60	40	40	40	40		40
Picking rate (units/hr)	40	40	40	40	40		40
Hourly labor rate ($/hr)	200	222	190	192	185		180
No of forklifts (count)	15	18	18	18	17.50		18
No of trucks (count)	4	6	10	4	10		4
	2	3	8	6	6		3

FORMULAS

1. Service cost = cost to order+ cost of del. into store + cost of merchandising
2.
3. Dist.cost=(Dist to zone's center)*(cost/mile)
4. Replen cost = avgcost/unit/lane
5. Avg Utilization = (Total labor hours) / (Total available hrs)
6.
7.
8.
9.
10.

SYMBOLOGY

- MANUFACTURING WITH CROSSDOCK
- OUTSIDE SUPPLIER
- DISTRIBUTION CENTER
- RAW MATERIAL SUPPLIER
- FINISHED GOODS
- RAW MATERIAL

© Copyright 2007. CHANDRA NATARAJAN AND RICHARD MUTHER & ASSOCIATES -775

May be reproduced for in-company use provided original source is not deleted.

BASELINE VALIDATION WORKSHEET

Company: KMG Inc
By: LHH, SN, KP
Date: 8/6
Project Name: Lagrange Operations strategy
With: Sales, Ops, SC, Finance
Sheet: 1 of 1

VARIABLES	UOM	MODEL RESULTS	ACTUALS	VARIANCE	OK	NOTES
DEMAND						
1. Total units produced by Marietta GA	Units	12,870,000	12,500,000	(370,000)	✓	1.
2. Total units produced by Columbia SC	Units	10,827,000	11,000,000	173,000	✓	2.
3. Total units produced by Jacksonville FL	Units	8,230,000	8,500,000	270,000	✓	3.
4. Total units handled by Albany GA	Units	2,100,000	2,120,000	20,000	✓	
5. Total units handled by Lagrange GA	Units	1,200,000	1,160,000	(40,000)	✓	4.
6. Total units handled by Augusta GA	Units	2,300,000	2,390,000	90,000	✓	5.
7. Total units handled by Savannah GA	Units	2,200,000	2,180,000	(20,000)	✓	
8. Total units received from outside suppliers	Units	4,200,000	4,200,000	0		
9.						
10.						
COST						
1. Warehousing operating cost	$	$2,220,000	$2,285,000	$65,000	✓	6.
2. Warehouse indirect labor cost	$	$1,080,000	$1,100,000	$20,000	✓	7.
3. Transportation (replenishment) cost	$	$3,210,000	$3,260,000	$50,000	✓	8.
4. Inventory cost	$	$477,000	$480,000	$3,000	✓	
5. Distribution operating cost	$	$6,200,000	$6,118,000	($82,000)	✓	9.
6. Distribution labor cost	$	$12,400,000	$12,620,100	$220,100	✓	10.
7. Infrastructure cost	$	$420,000	$422,000	$2,000	✓	
8.						
9.						
10.						

CONSTRAINTS

		OK
1.	Customer demand must be met.	✓
2.	Products will flow as per the current sourcing set up.	✓
3.	All lanes can only move full truck loads of product.	✓
4.	The storage capacity at each distribution center is fixed.	✓
5.	Each customer store must be served by only one DC	✓
6.	Total inventory at any point in any DC will be modeled as avg. inventory + safety stock.	✓
7.	Customers will be served at fixed frequencies to accurately depict service & dist. costs.	✓
8.	Distribution trucks will carry only the needed demand (based on route plan)	✓
9.	Warehouse labor will be available for 80 hrs each week	✓
10.	Labor hours may not exceed fixed resources (e.g. LaGrange has 6 pickers)	✓

NOTES / EXPLANATION:

1. There was a line maintenance at Marietta, GA and Marietta was shut down for 1 week.
2. Columbia, SC produced part of the Marietta shut down units.
3. Jacksonville, FL produced part of the Marietta shut down units.
4. LaGrange GA directly handled fewer cases because of routing re-alignment to Marietta.
5. Augusta GA directly sourced products from outside supplier when Marietta line was shut down.
6. Warehouse operating cost variance was due to the increased overtime which was run at Jacksonville, Columbia, Augusta
7. Warehouse indirect labor cost variance was due to the model variance and since it was within 2%, it was accepted by validation team
8. Transportation cost variance was $50,000 because of the common carrier impact that the KMG sometimes uses during peak season
9. The distribution fixed cost variance was due to the new trucks that added to the fleet.
10. The distribution labor cost variance is attributed to the model variance and since the variance being within 2% , it was considered ok.

© Copyright 2007. CHANDRA NATARAJAN AND RICHARD MUTHER & ASSOCIATES - 777   May be reproduced for in-company use provided original source is not deleted.

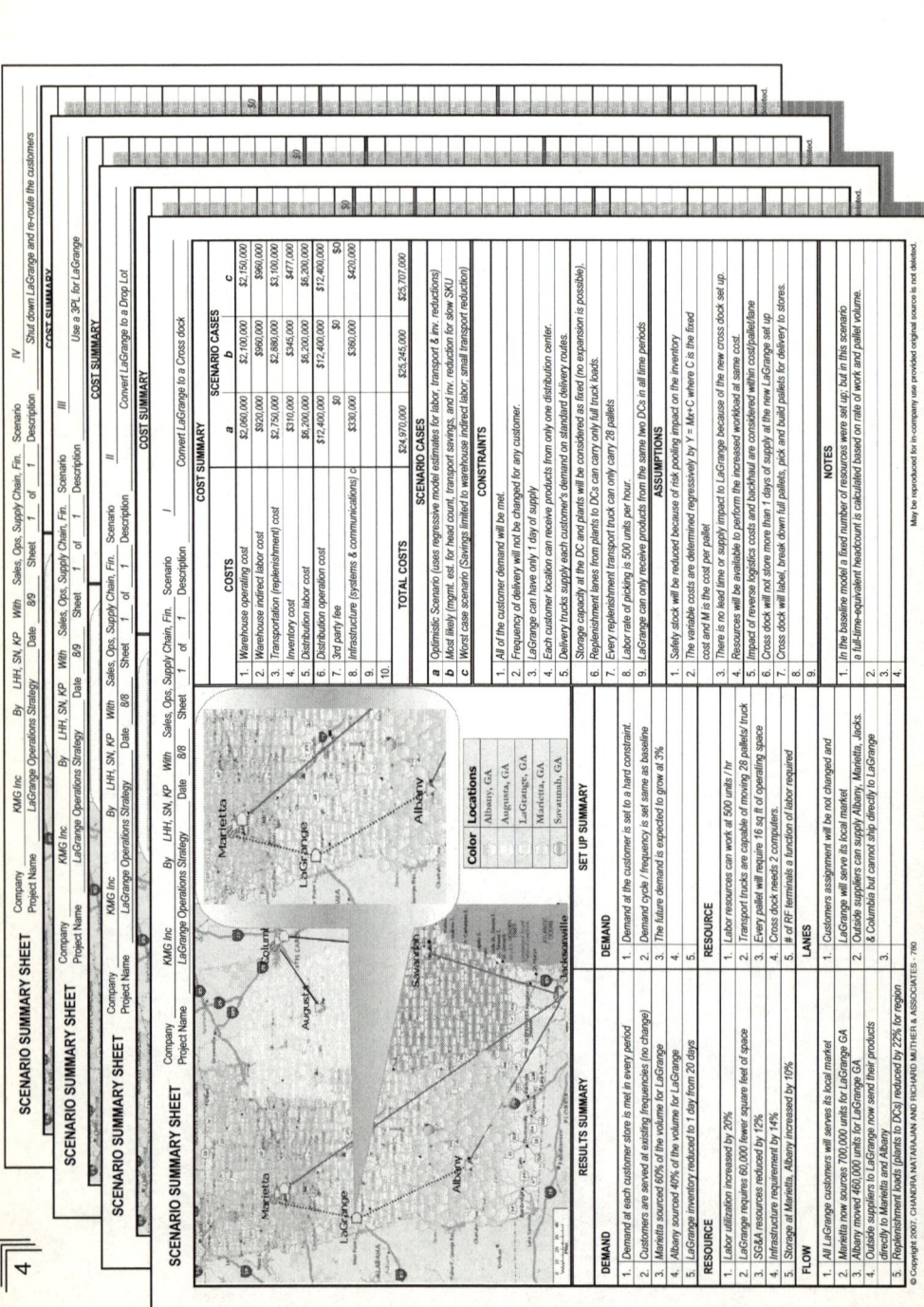

ALTERNATIVES ANALYSIS WORKSHEET — 5

Company: **KMG**
By: **LHH, SN, KP**
Date: **8/15**
Project Name: **LaGrange Operations Strategy**
With: **Sales, Ops, SC, Finance**
Sheet: **1** of **1**

COST ANALYSIS

By: LHH, SN, KP With: Sales, Ops, SC, Finance

COST SUMMARY

#	Item	BASELINE	Alt. I	Alt. II	Alt. III	Alt. IV
1	Warehouse operating cost	$2,220,000	$2,100,000	$1,930,000	$1,760,000	$1,760,000
2	Warehouse indirect labor cost	$1,080,000	$960,000	$920,000	$780,000	$780,000
3	Transportation (replenishment) cost	$3,210,000	$2,880,000	$2,970,000	$2,980,000	$2,320,000
4	Inventory cost	$477,000	$345,000	$330,000	$330,000	$300,000
5	Distribution labor cost	$6,200,000	$6,200,000	$6,200,000	$6,400,000	$6,920,000
6	Distribution operation cost	$12,400,000	$12,400,000	$12,400,000	$12,700,000	$13,190,000
7	Third party fee (1)	$0	$0	$0	$300,000	$0
8	Infrastructure (systems & communications) cost	$420,000	$360,000	$320,000	$310,000	$300,000
9						
10						
	TOTAL ANNUALIZED COSTS	$26,007,000	$25,245,000	$25,070,000	$25,560,000	$25,570,000
	ANNUALIZED SAVINGS OVER BASELINE		-$762,000	-$937,000	-$447,000	-$437,000

INTANGIBLE ANALYSIS

Weights by: LH, SN   Ratings by: MM, LH, DW, TR, KP, TF, JJE   Tally by: LH, SNN

#	FACTOR / CONSIDERATION	WT.	BASELINE	Alt. I	Alt. II	Alt. III	Alt. IV
1	Impact on corporate image	1		E / 3	E / 3	U / 0	X / 0
2	Impact on customer service	8		E / 24	I / 16	O / 8	U / 0
3	Process required to enable change (2)	9		A / 36	O / 9	O / 9	A / 36
4	Impact on strategy (3)	10		A / 40	I / 20	O / 10	E / 30
5	Impact on local community	5		E / 15	O / 5	I / 10	I / 10
6	Availability of labor in Marietta & Albany	6		O / 6	U / 0	A / 24	U / 0
7	Local management acceptance (4)	7		E / 21	I / 14	U / 0	U / 0
8							
9							
10							
	TOTAL INTANGIBLES EFFECT			145	67	61	76

NOTES

1. Outsourcing fee paid to MTX, a third party logistics company.
2. KMG doesn't have processes or systems to support the drop lot operations.
3. KMG wants to operate with fewer inventory carrying locations.
4. KMG wants a plan that is readily accepted by local managers.

SCENARIO NAME

- **BASELINE**: Lagrange will be a DC and will be operating as is
- **I**: Convert LaGrange to a Cross Dock
- **II**: Convert LaGrange to a Drop lot
- **III**: Use a 3PL for LaGrange
- **IV**: Shut down LaGrange and re-route the customers

EVALUATING DESCRIPTION

A	Almost Perfect	O	Ordinary Results
E	Especially Good	U	Unimportant Results
I	Important Results	X	Not Acceptable

Values: A = 4, E = 3, I = 2, O = 1, U = 0, X = -1

COST DIFFERENCE FROM BASELINE

SAVINGS	INCREASED COSTS
-26% to -50%	0% to 5%
-16% to -25%	6% to 15%
-6% to -15%	16% to 25%
0% to -5%	26% to 50%

SUMMARY

1. Alt. II Droplot is best in terms of cost.
2. From the intangible perspective Alt. I crossdock is best.
3. All factors considered, the best strategy is to operate LaGrange as a cross dock.

© Copyright 2007. CHANDRA NATARAJAN AND RICHARD MUTHER & ASSOCIATES - 178

DETAIL AND DO WORKSHEET

Covering	Sales, Operations, Supply Chain, Finance
Distribution	MM, LH, DW, TR, KP, TF, JJB, PI, PV
Company	KMG
Status as of	9/7
Reported by	SN
Project	LaGrange Operations Strategy
Project Description	Identify best strategy for operating LaGrange
Date	9/7
By	LHH, SN, KP
Sheet	1 of 1

Task/Proj. No.	Work to do; Action to take	Resp. of	8/20-8/24	8/27-8/31	9/3-9/7	9/10-9/14	9/17-9/21	Further Schedule
1.	Set up LaGrange as a crossdock receiving location	PI	■					
2.	Identify and tag inventory to be moved from LaGrange	SN	■	■				
3.	Communicate the sourcing changes for depleting the inventory at LaGrange	MM		■				
4.	Determine the staffing plan for LaGrange, Marietta, Albany	LH		■				
5.	Personnel issues and HR communications (severance plans)	DW		■				
6.	Move equipment and inventory from LaGrange	TR		■				
7.	Set up the processes and communications to support cross docking	TR		■				
8.	Set up Marietta and Albany as crossdock suppliers (builders of pallets)	KP		■				
9.	Change the sourcing plan & set up the SKU's	TF		■				
10.	Forecast the replenishment requirements for the cross docks	JJB			■			
11.	Communicate with sales team the changes to operations	JJB			■			
12.	Pilot the operations on two routes	PV			■			
13.	Understand the lessons learned	SN				■		
14.	Update the operations and deploy the solution across all routes(LaGrange)	SN				■	▼	
15.								

POST IMPLEMENTATION AUDIT

No.	COST SUMMARY	UOM	Projected Savings	Actual Savings	Variance	Notes
1.	Warehouse operating cost	$	$120,000	$90,000	($30,000)	1.
2.	Warehouse labor cost	$	$200,000	$180,000	($20,000)	2.
3.	Transportation (replenishment) cost	$	$330,000	$360,000	$30,000	3.
4.	Inventory cost	$	$132,000	$160,000	$28,000	4.
5.	Distribution labor cost	$	$0	$0	$0	
6.	Distribution operation cost	$	$0	$0	$0	
7.	3rd Party fee	$	$0	$0	$0	
8.	Infrastructure cost	$	$60,000	$42,000	($18,000)	5.
9.	Real estate savings	$	$0	$350,000	$350,000	6.
10.						

No.	NOTES AND EXPLANATION AND LESSONS LEARNED
1.	Warehouse required one additional supervisor to monitor the cross dock operations.
2.	Labor savings was reduced by $20,000 because Marietta and Jacksonville had to hire additional labor and there was a learning curve to get pickers to same level of picking.
3.	Transportation savings was increased by $30,000 because a common carrier was available to move product to Jacksonville at a cheaper cost.
4.	Inventory savings increased by $28,000 because many slow moving LaGrange SKU's were consolidated at Marietta.
5.	Additional RF terminals were required to support the cross dock operations.
6.	Space not needed in LaGrange warehouse rented to neighboring tenant.
7.	
8.	
9.	

☐ Date work scheduled to start
— Amount of work done
☐ Date work scheduled to finish
▼ Reporting indicator
☐ Total time scheduled for work (Each vertical period represents one unit of time)

© Copyright 2007. CHANDRA NATARAJAN AND RICHARD MUTHER & ASSOCIATES - 768

May be reproduced for in-company use provided original source is not deleted.

CASE STUDY 3 MONTAGE

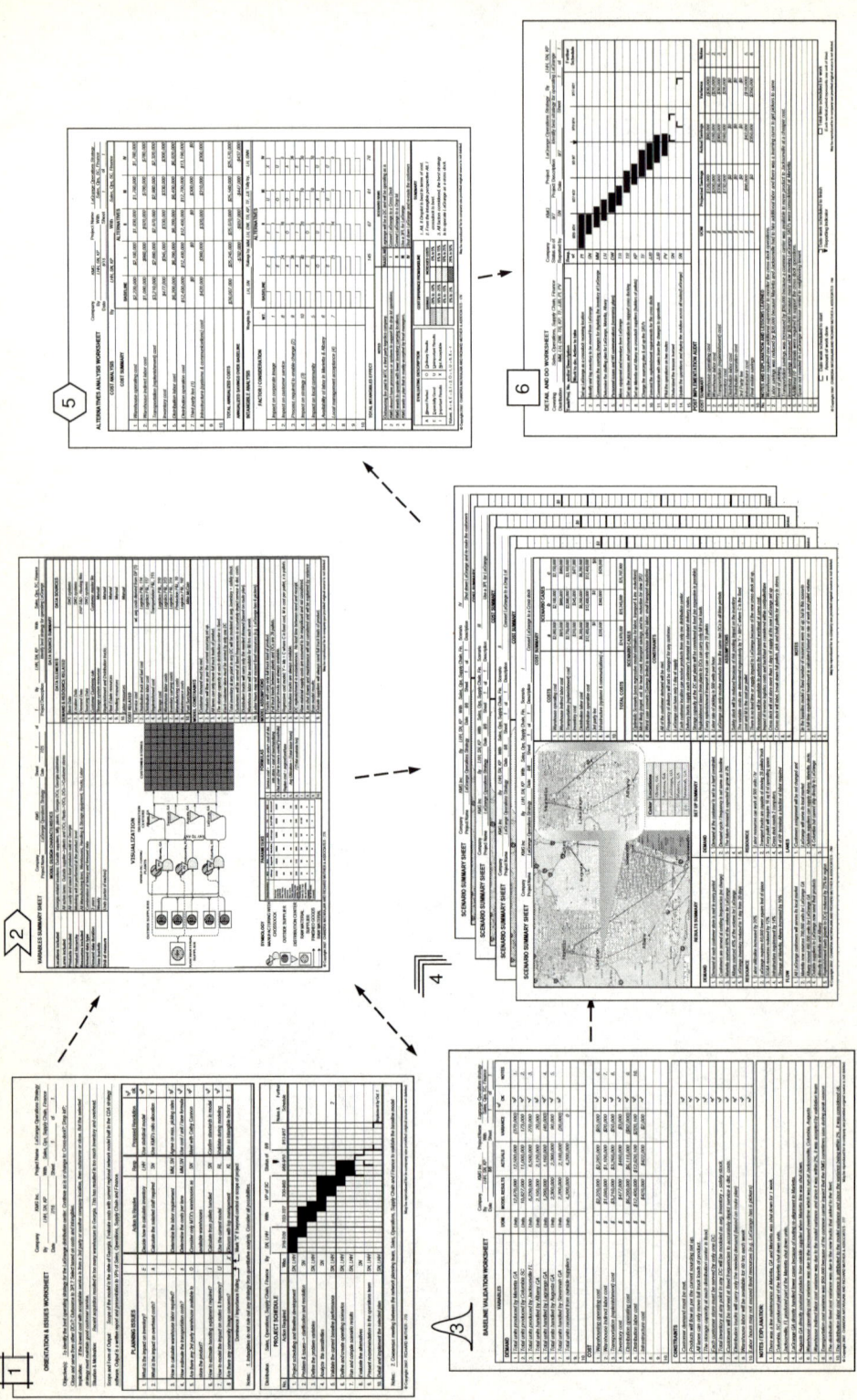

WORKING FORMS

Following this sheet are copies of the printed forms most frequently used for short form Systematic Network Planning. They may be removed and copied for use when planning your next project. If you will remove the form by constant-angle tear out tuck it back in the binding for convenient filling. All the forms can be used for simulation project along with network analysis and may be used in more than one ways in the real life projects.

You may reproduce the copies of these forms for your own use, provide you recognize their source and hold their use with the copyright restrictions covering this book.

Chapter	Form No.	Form Name or Description
1.	RMA - 773	*Orientation and Issues Worksheet*
2.	RMA - 775	*Variables Summary Worksheet*
3.	RMA - 777	*Baseline Validation Worksheet*
4.	RMA - 780	*Scenarios Summary Worksheet*
5.	RMA - 178	*Alternatives Analysis Worksheet*
6.	RMA - 768	*Detail and Do Worksheet*

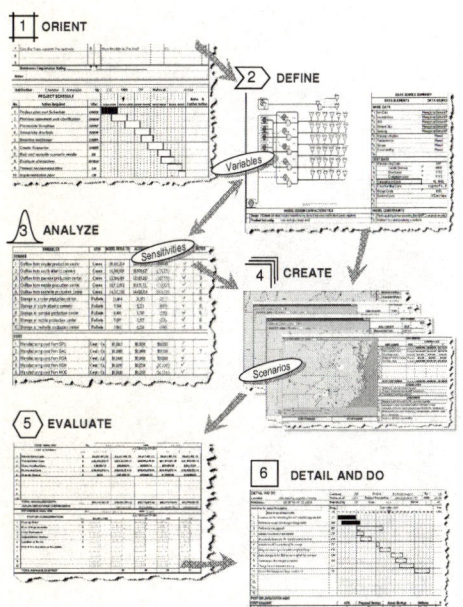

Of course, you may use your own forms if they suit your project better or are more-or-less standard in your organization.

Of course you can adapt or redesign these forms for your specific purposes.

Of course you can put them on your computer

Of course you may not choose to use pre-designed forms on each step, or even use them at all.

We have tried to be helpful. In our experience, a well-structured form organizes your efforts; the project and the process are both more easily understood; the "solution" often is much more apparent; and when filled-in a form can serve as a powerful example for helping team members understand what their thinking process might/will "look" like.

ORIENTATION & ISSUES WORKSHEET

Company: _____ Project Name: _____
By: _____ With: _____
Date: _____ Sheet: _____ of _____

Objective(s): _____

Implication: _____

Situation & Motivation: _____

Scope and Form of Output: _____

PLANNING ISSUES			Action to Resolve	Resp	Proposed Resolution	✓ ok
1.						
2.						
3.						
4.						
5.						
6.						
7.						
8.						
9.						

Dominance / Importance Rating ↑ ↑ Mark "X" if beyond control or scope of project

Notes: _____

Distribution: _____ By: _____ With: _____ Status of: _____

PROJECT SCHEDULE

No.	Action Required	Who										Notes & Further Schedule
1.												
2.												
3.												
4.												
5.												
6.												
7.												
8.												
9.												
10.												

Notes: _____

© Copyright 2007. RICHARD MUTHER - 773

May be reproduced for in-company use provided original source is not deleted.

VARIABLES SUMMARY SHEET

Company _____ Sheet _____ of _____ By _____ With _____
Project Name _____ Date _____ Project Description _____

MODEL DESIGN CHARACTERISTICS

Locations included	
Lanes included	
Products included	
Product hierarchy	
Resources included	
Demand data type	
Demand data duration	
Time buckets	
Unit of measure	

VISUALIZATION

DATA SOURCE SUMMARY

DATA ELEMENTS | DATA SOURCES

DEMAND & RESOURCE RELATED
1.
2.
3.
4.
5.
6.
7.
8.
9.
10.

COST RELATED
1.
2.
3.
4.
5.
6.
7.
8.
9.
10.

MODEL CONSTRAINTS
1.
2.
3.
4.
5.
6.
7.
8.
9.
10.

MODEL ASSUMPTIONS
1.
2.
3.
4.
5.
6.
7.
8.
9.
10.

PARAMETERS | FORMULAS

1.
2.
3.
4.
5.
6.
7.
8.
9.
10.

SYMBOLOGY

- MANUFACTURING WITH CROSSDOCK
- OUTSIDE SUPPLIER
- DISTRIBUTION CENTER
- RAW MATERIAL SUPPLIER
- FINISHED GOODS
- RAW MATERIAL

© Copyright 2007. CHANDRA NATARAJAN AND RICHARD MUTHER & ASSOCIATES - 775

May be reproduced for in-company use provided original source is not deleted.

BASELINE VALIDATION WORKSHEET

Company: _____ Project Name: _____
By: _____ With: _____
Date: _____ Sheet: _____ of _____

VARIABLES	UOM	MODEL RESULTS	ACTUALS	VARIANCE	✓ OK	NOTES
DEMAND						
1.						
2.						
3.						
4.						
5.						
6.						
7.						
8.						
9.						
10.						
COST						
1.						
2.						
3.						
4.						
5.						
6.						
7.						
8.						
9.						
10.						

CONSTRAINTS
1.
2.
3.
4.
5.
6.
7.
8.
9.
10.

NOTES / EXPLANATION:
1.
2.
3.
4.
5.
6.
7.
8.
9.
10.

© Copyright 2007. CHANDRA NATARAJAN AND RICHARD MUTHER & ASSOCIATES - 777

May be reproduced for in-company use provided original source is not deleted.

SCENARIO SUMMARY SHEET

Company _____ By _____ With _____ Scenario _____
Project Name _____ Date _____ Sheet _____ of _____ Description _____

VISUALIZATION

COST SUMMARY

COSTS	SCENARIO CASES		
	a	b	c
1.			
2.			
3.			
4.			
5.			
6.			
7.			
8.			
9.			
10.			
TOTAL COSTS			

SCENARIO CASES

a	
b	
c	

CONSTRAINTS

1.
2.
3.
4.
5.
6.
7.
8.
9.
10.

ASSUMPTIONS

1.
2.
3.
4.
5.
6.
7.
8.
9.
10.

NOTES

1.
2.
3.
4.
5.

SET UP SUMMARY

DEMAND
1.
2.
3.
4.
5.

RESOURCE
1.
2.
3.
4.
5.

LANES
1.
2.
3.
4.
5.
6.

RESULTS SUMMARY

DEMAND
1.
2.
3.
4.
5.

RESOURCE
1.
2.
3.
4.
5.

FLOW
1.
2.
3.
4.
5.
6.

© Copyright 2007, CHANDRA NATARAJAN AND RICHARD MUTHER & ASSOCIATES - 780   May be reproduced for in-company use provided original source is not deleted.

ALTERNATIVES ANALYSIS WORKSHEET

Company _____ Project Name _____
By _____ With _____
Date _____ Sheet _____ of _____

COST ANALYSIS By _____ With _____

COST SUMMARY	BASELINE	ALTERNATIVES			
		I	II	III	IV
1.					
2.					
3.					
4.					
5.					
6.					
7.					
8.					
9.					
10.					
TOTAL ANNUALIZED COSTS					
ANNUALIZED SAVINGS OVER BASELINE					

INTANGIBLE ANALYSIS Weights by: _____ Ratings by: _____ Tally by: _____

FACTOR / CONSIDERATION	WT.	BASELINE	ALTERNATIVES			
			I	II	III	IV
1.						
2.						
3.						
4.						
5.						
6.						
7.						
8.						
9.						
10.						
TOTAL INTANGIBLES EFFECT						

NOTES
1.
2.
3.
4.
5.

SCENARIO NAME
BASELINE	
I	
II	
III	
IV	

EVALUATING DESCRIPTION
A	Almost Perfect	O	Ordinary Results
E	Especially Good	U	Unimportant Results
I	Important Results	X	Not Acceptable

Values: A = 4, E = 3, I = 2, O = 1, U = 0, X = -1

COST DIFFERENCE FROM BASELINE
SAVINGS	INCREASED COSTS
-26% to -50%	0% to 5%
-16% to -25%	6% to 15%
-6% to -15%	16% to 25%
0% to -5%	26% to 50%

SUMMARY

DETAIL AND DO WORKSHEET

Covering _____
Distribution _____

Company _____
Status as of _____
Reported by _____

Project _____
Project Description _____
Date _____

Sheet ___ of ___
By _____
With _____

Task/Proj. No. and/or Description		Resp. of						Further Schedule
	Work to do; Action to take							
1.								
2.								
3.								
4.								
5.								
6.								
7.								
8.								
9.								
10.								
11.								
12.								
13.								
14.								
15.								

POST IMPLEMENTATION AUDIT

COST SUMMARY

	UOM	Projected	Actual	Variance	Notes
1.					
2.					
3.					
4.					
5.					
6.					
7.					
8.					
9.					
10.					

NOTES AND EXPLANATION AND LESSONS LEARNED

No.	
1.	
2.	
3.	
4.	
5.	
6.	
7.	
8.	
9.	

☐ Date work scheduled to start
▬ Amount of work done

☐ Date work scheduled to finish
▼ Reporting indicator

☐ Total time scheduled for work
(Each vertical period represents one unit of time)

May be reproduced for in-company use provided original source is not deleted.

© Copyright 2007, CHANDRA NATARAJAN AND RICHARD MUTHER & ASSOCIATES . 768

SYNOPSIS OF FULL SYSTEMATIC NETWORK PLANNING (SNP)

Systematic Network Planning (SNP) is an organized, universally-applicable approach to any network planning project that employs network modeling software. SNP consists of:

- A Framework of Phases
- A Pattern of Procedures
- A Set of Conventions

These are pictured in the SNP Capsule Summary on the next page.

The Four Phases of Network Planning

As each project runs its course, it passes through the four phases of network planning:

> **Phase I** is that of **Orientation**. Here we must decide the scope of the project and the network to be modeled. We also establish the network planning objectives, the desired outputs and a schedule for the project.
>
> **Phase II** is that of the **Overall Network Plan**. Here we decide for groups or families of items and for regions of demand the number of facilities, their general locations and capacities. The Overall Plan is typically developed with just a few critical constraints on model solutions.
>
> **Phase III** is the preparation of **Detailed Network Plans** for specific items, transportation methods, storage and production capacities, and with additional constraints. Phase III models are typically detailed instances or sub-sets of the higher-level or aggregate model used in Phase II – with a detailed plan prepared for each region, family of items, or set of routes.
>
> **Phase IV** is the **Implementation**. This involves both planning and making the specific changes necessary to implement the detailed network plans.

These phases come in sequence and for best results they should overlap each other.

Basic Input Data for Network Planning

SNP recognizes basic information required for network planning. It keys them to the "alphabet of physical operations planning" – P Q R S T. Practically every network plan relies on these elements as a basis for its planning:

> P – Products, Materials or items being moved through the network.
>
> Q – Quantities or volumes of products, materials or items – both in terms of demand flow rates and target inventory levels.
>
> R – Routing or process, i.e. the operations performed, their sequence and equipment required.
>
> S – Services or supporting activities which back up the processing, warehousing, and transportation operations.
>
> T – Time as it relates to P Q R S and the scheduling of the project itself. This can include peaks and seasonality, hours and days of operation, cut-off times…

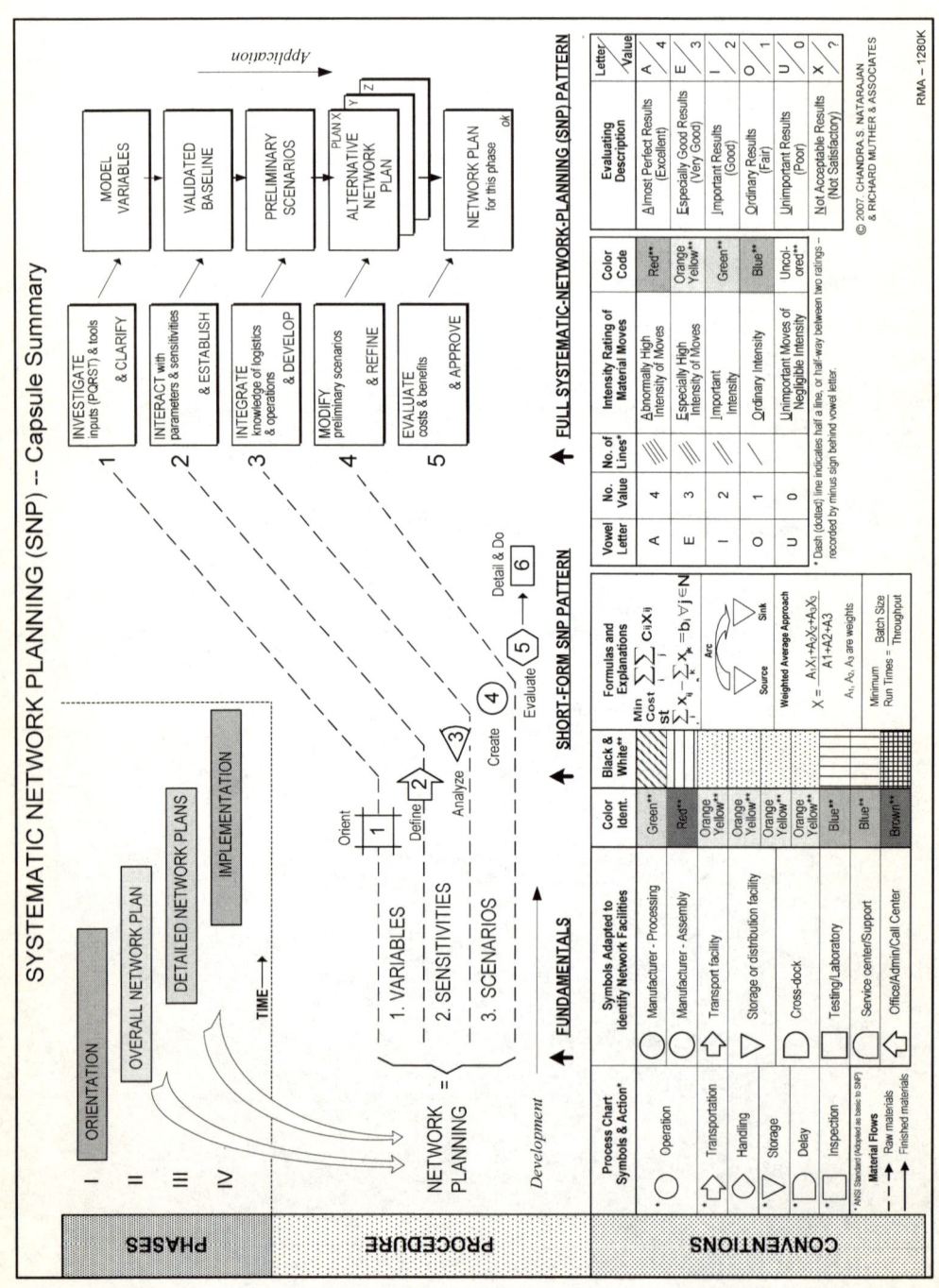

The Pattern of Procedures, Overall Network Plan

The overall network plan begins with gathering and investigation of input data (PQRST), set-up of modeling tools and a clarification of the variables to be modeled. We visualize the network and formulate the model in terms of demand, costs, constraints, and process parameters. Once our data has been scrubbed and entered into the model, we run and revise it until we have eliminated infeasible results. Next, we test the sensitivities of variables to changes in parameters and establish a baseline model that replicates the actual performance of our current network. Using our knowledge of logistics and operations, we then develop several preliminary scenarios, each representing and alternative network plan. These are modified and refined until we finally end up with two, three, four, or even five alternative network proposals. Each of these plans will work; each has value. At this point, projected costs are reviewed together with an analysis of intangible factors. This evaluation results in a choice of one network plan.

The Pattern of Procedures, Detailed Network Plans

The next phase is that of preparing detailed network plans. This involves item-, location- and route-specific decisions within the overall network plan. Phase II overlaps Phase III. This means that before actually finishing the overall network plan, certain details will have to be analyzed and worked out. For example, a tentative facility location may be found in Phase II by modeling with aggregated customer demand. But to confirm this location, the planners may want to run a more detailed analysis using less aggregated or even specific customer locations. A make-buy analysis in Phase II may identify an entire family or set of items that should be bought. But detailed analysis may be needed to confirm specific items.

Note also that a detailed network plan must be made for each of the regions and/or subsets of items and routes in the overall plan. This some adjustment may have to be made to the overall model and plan.

The same Pattern of Procedures is used in both Phases II and III. Again, for each sub-network, or subset of items or routes, we end up with several alternative plans. This leads to an evaluation to select the best.

A Set of Conventions

A set of conventions aids in the planning and understanding. These consist of:

> Symbols for representing operations and types of network facilities in diagrammatic form.
>
> Colors and black and white shading patterns for representing types of network facilities.
>
> Formulas for common modeling objectives, representations and approximations.
>
> Vowel letters, line ratings, and colors for representing intensities of material flow in diagrammatic form.
>
> Vowel letter ratings for evaluating the performance of network plans.

Conclusions

By approaching problems in four overlapping phases, SNP helps planners to tackle networks of any size and scope. The necessary modeling software must be available, of course. The addition of SNP's logical and simple "thoughtware" can save many hours and avoid many delays, frustrations and misunderstandings. It puts into proper relation the many procedures and techniques that can be used in network planning. SNP was developed from actual planning projects and has been proven in actual use. When followed, it can bring sequence and clarity to what on each project seems a "different" problem. And while every network project *is* different – for no two networks are exactly alike -- SNP provides a standard framework, procedures and conventions useful for planning any network.

Differences between full SNP and the Simplified SNP method described in this booklet are summarized on the inside of the back cover.

DIFFERENCES BETWEEN <u>SIMPLIFIED SNP</u> AND FULL <u>SYSTEMATIC NETWORK PLANNING – SNP</u>

Simplified S.N.P.	Systematic Network Planning -- SNP
• Six steps.	• Four phases, with a five-section pattern that repeats in Phases II and III.

• Suited to small, simple projects, especially when:
   – Analysis is focused on one or a few products,
   – Transportation costs are dominant, and
   – The number of possible locations is small and they lie in a limited geographic region.
	• Can handle any size of project:
	   – Analysis involves many products,
	   – With significant differences in fixed and operating costs, in addition to transportation, and
	   – Many locations are possible in several dispersed regions.

• Easy to learn (less than one day). Highly suited to self-directed teams.	• User should understand the value of the full planning methodology.
• May be applied for detailed or local network planning within a larger project using full SNP.	• Full SNP projects may use Simplified SNP for Phase III sub-projects.
• For applying and revising previously developed network models.	• For initial development of new and large-scale models.

• Typical planning problems:
   – Change in inventory policy
   – Adding capacity
   – Adding or closing branch
   – Compare transportation modes
   – Where to make a new product (among existing sites)
   – Change in lot size
   – Make or buy
   – Overtime or Pre-Build
	• Typical planning problems:
	   – Where to locate new facility
	   – Capacity planning across regions and for multiple product lines
	   – Assignments of multiple products to multiple facilities
	   – Sourcing analyses for multiple sites and products
	   – Long range operations strategies

SIMPLIFIED S.N.P SUMMARY

1. ORIENT THE PROJECT

a. Identify the project, its objectives and scope.
b. Understand the problem and the elements to be modeled.
c. Document and rate the planning issues.
d. Make a plan for the network analysis.

Key Document: Orientations and Issues Worksheet
Output: The project understood and scheduled.

2. DEFINE THE VARIABLES

a. Define the model variables.
b. Visualize the network.
c. Gather inputs: demand, costs, constraints, process parameters.
d. List all assumptions and write the formulas to be used.
e. Gather, scrub and format the data.
f. Formulate the model and enter data to meet model requirements.

Key Document: Variables Summary Sheet
Output: Network model and data ready to run.

3. ANALYZE THE SENSITIVITIES

a. Run the model.
b. Identify infeasibilities, troubleshoot and create a model free from infeasibilities.
c. Run the model again to replicate current network performance and measure variance.
d. Adjust the model until variance from current performance is acceptable.
e. Summarize baseline model and current network statistics.

Key Document: Baseline Validation Worksheet
Output: Validated baseline model that replicates current performance.

4. CREATE SCENARIOS

a. Identify potential scenarios that model the problem elements.
b. Add or remove variables from the baseline to represent each potential scenario.
c. Collect summary statistics for each alternative plan and document the results.

Key Document: Scenario Summary Sheet
Output: Model results for alternative network plans.

5. EVALUATE THE ALTERNATIVES

a. Prepare an Alternatives Analysis Worksheet.
b. For costs, list the positive and negative cash flows; add and annualize them.
c. For intangibles, list the factors, assign weights, rate each plan; extend and downtotal.
d. Compare and select the most preferred.

Key Document: Alternatives Analysis Worksheet
Output: The selected network plan.

6. DETAIL AND DO

a. List the actions required to implement the selected network plan.
b. Plan and schedule the implementation.
c. Periodically post performance against the scheduled task and take appropriate actions.
d. Prepare the post-implementation audit and record the lessons learned.

Key Document: Detail And Do Worksheet
Output: Implemented plan and audited results.

© 2007. Chandra S. Natarajan and Richard Muther & Associates